Environment and Law

Environmental problems such as global warming and pollution lie at the heart of the public agenda in the twenty-first century. To be effective in tackling these, environmental law must be prepared to adopt practical strategies and techniques from the fields of economics, science, ethics and politics, to further an understanding of the proper form and content of environmental law itself.

Environment and Law initially describes and explains law and legal systems, the concept of the environment, sources of environmental law and some of the techniques used in environmental law. It then examines some of the major connections between law and the disciplines of economics, science, ethics, and politics. Some of the issues discussed are:

- how economic instruments can offer alternatives and supplements to traditional 'command and control' forms of environmental regulation;
- the role of science in the resolution of environmental law disputes;
- the response of environmental law to the rise in theories of environmental ethics;
- and the kinds of political entities that are most conducive to high standards of environmental protection.

Environment and Law is a concise introduction for students with little or no legal background to the role of law in environmental protection. It offers a greater understanding of international and national environmental law and has case studies from all over the world, including examples from UK, US and Australian law. Chapter summaries, annotated further reading, a glossary of legal terms, a list of legal cases and their abbreviations are also included.

David Wilkinson is Lecturer in Environmental Law at the School of Politics, International Relations and the Environment, Keele University.

Routledge Introductions to Environment Series
Published and Forthcoming Titles

Titles under Series Editors:
Rita Gardner and A.M. Mannion

Environmental Science texts

Atmospheric Processes and Systems
Natural Environmental Change
Biodiversity and Conservation
Ecosystems
Environmental Biology
Using Statistics to Understand the Environment
Coastal Systems
Environmental Physics

Forthcoming:
Environmental Chemistry (September 2002)

Titles under Series Editor:
David Pepper

Environment and Society texts

Environment and Philosophy
Environment and Social Theory
Energy, Society and Environment
Environment and Tourism
Gender and Environment
Environment and Business
Environment and Politics (2nd edition)
Environment and Law

Representing the Environment (July 2002)
Environmental Movements (September 2002)
Environmental Policy (September 2002)
Environment and Society (December 2002)

Routledge Introductions to Environment Series

Environment and Law

David Wilkinson

London and New York

First published 2002 by Routledge
11 New Fetter Lane, London EC4P 4EE

Simultaneously published in the USA and Canada
by Routledge
29 West 35th Street, New York, NY 10001

Routledge is an imprint of the Taylor & Francis Group

Typeset in Times by Keystroke, Jacaranda Lodge, Wolverhampton
Printed and bound in Great Britain by Biddles Ltd, Guildford and King's Lynn

British Library Cataloguing in Publication Data
A catalogue record for this book is available from the British Library

Library of Congress Cataloging in Publication Data
A catalogue record for this book has been requested

ISBN 0–415–21567–6 (hbk)
ISBN 0–415–21568–4 (pbk)

Contents

Figures

Tables

Boxes

Series editor's preface *Environment and Society titles*

The modern environmentalist movement grew hugely in the last third of the twentieth century. It reflected popular and academic concerns about the local and global degradation of the physical environment which was increasingly being documented by scientists (and which is the subject of the companion series to this, *Environmental Science*). However it soon became clear that reversing such degradation was not merely a technical and managerial matter: merely knowing about environmental problems did not of itself guarantee that governments, businesses or individuals would do anything about them. It is now acknowledged that a critical understanding of socio-economic, political and cultural processes and structures is central in understanding environmental problems and establishing environmentally sustainable development. Hence the maturing of environmentalism has been marked by prolific scholarship in the social sciences and humanities, exploring the complexity of society–environment relationships.

Such scholarship has been reflected in a proliferation of associated courses at undergraduate level. Many are taught within the 'modular' or equivalent organisational frameworks which have been widely adopted in higher education. These frameworks offer the advantages of flexible undergraduate programmes, but they also mean that knowledge may become segmented, and student learning pathways may arrange knowledge segments in a variety of sequences – often reflecting the individual requirements and backgrounds of each student rather than more traditional discipline-bound ways of arranging learning.

The volumes in this *Environment and Society* series of textbooks mirror this higher educational context, increasingly encountered in the early twenty-first century. They provide short, topic-centred texts on social science and humanities subjects relevant to contemporary society–environment relations. Their content and approach reflect the fact that each will be read by students from various disciplinary backgrounds, taking in not only social sciences and humanities but others such as physical and natural sciences. Such a readership is not always familiar with the disciplinary background to a topic, neither are readers necessarily going on to further develop their interest in the topic. Additionally, they cannot all automatically be thought of as having reached a similar stage in their studies – they may be first-, second- or third-year students.

The authors and editors of this series are mainly established teachers in higher education. Finding that more traditional integrated environmental studies and specialised texts do not always meet their own students' requirements, they have often had to write course materials more appropriate to the needs of the flexible undergraduate programme. Many of the volumes in this series represent in modified form the fruits of such labours, which all students can now share.

Much of the integrity and distinctiveness of the *Environment and Society* titles derives from their characteristic approach. To achieve the right mix of flexibility, breadth and depth, each volume is designed to create maximum accessibility to readers from a variety of backgrounds and attainment. Each leads into its topic by giving some necessary basic grounding, and leaves it usually by pointing towards areas for further potential development and study. There is introduction to the real-world context of the text's main topic, and to the basic concepts and questions in social sciences/humanities which are most relevant. At the core of the text is some exploration of the main issues. Although limitations are imposed here by the need to retain a book length and format affordable to students, some care is taken to indicate how the themes and issues presented may become more complicated, and to refer to the cognate issues and concepts that would need to be explored to gain deeper understanding. Annotated reading lists, case studies, overview diagrams, summary charts and self-check questions and exercises are among the pedagogic devices which we try to encourage our authors to use, to maximise the 'student friendliness' of these books.

Hence we hope that these concise volumes provide sufficient depth to maintain the interest of students with relevant backgrounds. At the same time, we try to ensure that they sketch out basic concepts and map their territory in a stimulating and approachable way for students to whom the whole area is new. Hopefully, the list of *Environment and Society* titles will provide modular and other students with an unparalleled range of perspectives on society–environment problems: one which should also be useful to students at both postgraduate and pre-higher education levels.

David Pepper

May 2000

Series International Advisory Board

Author's preface

In this book I have attempted to provide a study of environment and law that is rather more systematic than those commonly to be found in other textbooks and casebooks on these subjects. Rather than discuss environmental law with the primary goal of relaying legal information to the reader, I have sought to explore the subject's structural aspects. What *is* environmental law? What *does it do?* What are its *underlying* principles, values and techniques? How does it *fit* and *relate* to the other key disciplines for environmental protection (ethics, economics, politics and science)? It is on these issues, which one might broadly term matters of 'environmental jurisprudence', that I have focused my thoughts.

Being an English author whose main knowledge lies in the English and European legal approaches to the environment might be thought of as something of a disadvantage when writing a book that seeks to cut across the boundaries of state and legal regime. Nevertheless, wherever possible I have included material from other jurisdictions, especially the United States, to show how similar or different approaches have been taken in other countries. I have also sought to avoid the use of complex legal terminology and, where this cannot be achieved, to explain legal terms as fully as possible. It is, therefore, my hope that those without any formal legal background will find the book accessible and interesting.

At times, the book may seem provocative, with the discussion ranging from the possible rights of aliens to the suggestions for the introduction of tradable birth permits. This is intentional. All who enter the wider debate concerning environmental protection will soon realise that one cannot formulate and defend solutions to environmental problems without addressing much more substantial philosophical questions: broadly, why should the environment matter, and what is the place of humans in the overall scheme of things? In attempting to answer these

Hotson v. *East Berkshire Health Authority* [1987] 3 WLR 232
Hunter v. *Canary Wharf Ltd.* [1997] 2 WLR 684
In re Wedgwood [1915] 1 Ch. 113
League Against Cruel Sports Ltd. v. *Scott* [1985] 2 All ER 589
Lemos v. *Kennedy Leigh Developments* (1961) 105 S.J. 178
Lord Advocate v. *Wemyss* [1900] AC 48
Lubbe and Others v. *Cape Plc* [2000] 4 All ER 268
McGhee v. *National Coal Board* [1973] 1 WLR 1
Maclaine Watson v. *Dept of Trade and Industry and related appeals* [1989] 3 All ER 523
Marquis of Granby v. *Bakewell UDC* (1923) 137 JP 105
Merlin v. *British Nuclear Fuels plc* [1990] 3 All ER 711
Monsanto plc v. *Tilly and others* (1999) 149 NLJ 1833, *The Times*, 30 November 1999
Moss v. *Christchurch R.D.C.* [1925] 2 KB 750
North Devon D.C. v. *West Coast Wind Farms* (1996) JPL 868
Pride of Derby and Derbyshire Angling Assoc. v. *British Celanese* [1953] Ch. 149
R. v. *Hertfordshire CC Ex p. Green Environmental Industries Ltd* [2000] Env LR 426
R. v. *Inspectorate of Pollution ex parte Greenpeace Limited* (No. 2), [1994] 4 All ER 329
R. v. *National Trust, ex parte Scott and Others* (1997), 16 July 1997, Queen's Bench Division, unreported
R. v. *Pollution Inspectorate ex parte Greenpeace* (No 2) [1994] 4 All ER 329
R. v. *Secretary of State for the Environment, ex parte Royal Society for the Protection of Birds* (Case 44/95)
R. v. *Secretary of State for the Environment and another, ex parte Standley and others* [1999] All ER (EC) 412 [1999] 2 CMLR 902
R. v. Secretary of State for the Environment, ex parte Rose Theatre Trust [1990] 1 QB 504
R. v. *Secretary of State for Foreign Affairs ex parte World Development Movement* [1995] 1 WLR 386
R. v. *Secretary of State for Trade and Industry, ex parte Duddridge and others* [1994] ELR 1
R. v. *Somerset County Council and ARC Southern Limited ex p Dixon*, 75 P&CR 175 [1997] JPL 1030
R. v. *Somerset County Council ex parte Fewings and others* [1995] 3 All ER 20
R. v. *Swale Borough Council, ex parte Royal Society for the Protection of Birds* [1991] JPL 39
R. v. *Environment Agency, ex parte Turnbull* (2000) unreported, Queen's Bench Division, 12 January 2000
Ransom & Daniels v. *Sir Robert McAlpine & Sons* [1971] K.I.R. 141

Re Celtic Extraction Ltd (in liquidation); *Re Bluestone Chemicals Ltd (in liquidation)* [1999] 4 All ER 684
Re Grove-Grady [1929] 1 Ch. 557
Re Shetland Salmon Farmers Association and Others [1991] SLT 166, 1990 SCLR 484
Reay and Hope v. *British Nuclear Fuels*, 5 *Water Law* 22
Redland Bricks Ltd v. *Morris* [1970] AC 652
Rickards v. *Lothian* [1912] AC 263
Rylands v. *Fletcher* (1868) L.R. 3 H.L. Cas 330
Salvin v. *North Brancepeth Coal and Coke Co.* [1874] 9 Ch. App 705
Scott-Whitehead v. *National Coal Board and Southern Water Authority*, 53 P&CR 263
Sedleigh-Denfield v. *O'Callaghan* [1940] AC 880
Shelfer v. *City of London Electric Lighting Co.* [1895] 1 Ch. D. 287
Singh v. *Observer Ltd.* [1989] 2 All ER 751
Southern Water Authority v. *Nature Conservancy Council* [1992] 3 All ER 481
St Helens Smelting v. *Tipping* (1865) 11 H.L. Cas 642
Sturges v. *Bridgman* (1879) 11 Ch. D. 852
Swan Fisheries Ltd. v. *Holberton* (1987), unreported
Tesco Stores Ltd. v. *Secretary of State* [1995] 2 All ER 134
Trendtex Trading Corporation v. *Central Bank of Nigeria* [1977] QB 529
Victoria Park Racing and Recreation Grounds Co Ltd v. *Taylor* (1937) 58 CLR 479
Wilsher v. *Essex Area Health Authority* [1988] 2 WLR 557

European cases

Association Greenpeace France and Others and Ministère de l'Agriculture et de la Pêche and Others (Case C-6/99)
Comitato di Coordinamento per la Defesa della Cava v. *Regione Lombardia* [1994] 1 ECR 1337
Commission v. *Denmark* [1988] ECR 4607
Commission v. *Germany* [1991] ECR 1-825
Commission v. *Ireland* (1996) Case C-392/96
Commission v. *Spain*, Case C-355/90 [1993] ECR I-4221
Danielsson v. *Commission*, Case T-219/95 R [1996] ECR II-3051
Eurostock Meat Marketing Ltd v. *Department of Agriculture for Northern Ireland* (Case C-477/98)
France v. *United Kingdom*, Case C-141/78 [1979] ECR 2923
Francovich v. *Italy* [1991] ECR 1-5357
Lopes Ostra v. *Spain* (1994) Ser. A No. 303C
McGinley and Egan v. *UK* (1998) *The Times*, 15 June 1998

Marleasing SA v. *La Commercial Internacional de Alimentacion SA* [1990] ECR 1-4135
Peralta, Case C-379/92 [1994] ECR I-3453
Powell and Rayner v. *UK* (1990) 12 EHRR 355
R. v. *Medicines Control Agency, ex parte Rhône-Poulenc Rorer Ltd and another* (Case C-94/98) [2000] All ER (EC) 46
Rewe v. *Hauptzollamt Kiel* [1981] ECR 1805
Safety Hi-Tech v. *S & T, Gianni Bettati* [1999] 2 CMLR 515
The Red Grouse Case [1990] I ECR 2143
World Wildlife Fund (WWF) EA and others v. *Autonome Provinz Bozen* [2000] 1 CMLR 149

US cases

Agins v. *City of Tiburon* (1980) 447 US 255
Aguida v. *Texaco* (1996) 845 F. Supp. 625
Allen v. *United States* (1984) 736 F.2d 1529, D.C. Cir. Ct. of Appeals
Amlon Metals, Inc. v. *FMC Corp.* (1991) 775 F. Supp. 668
Ayers v. *Township of Jackson* (1983) 189 N.J. Super. 561, 461 A.2d 184
Beanal v. *Freeport-Moran Inc.* (1997) US Dst Lexis 4767
Big Rock Mountain Corporation v. *Stearns Roger Corporation* (1968) 388 F.2d 165, 8th Cir.
Boomer v. *Atlantic Cement Company* (1970) 257 NE 2d 870
Chevron Chemical Company v. *Ferebee* (1984) 736 F.2d 1529, D.C. Cir. Ct. of Appeals
Chicago Steel and Pickling Co. v. *Citizens for a Better Environment*, 90 F.3d 1237
City of Philadelphia v. *New York* (1978) 437 US 617
Daubert v. *Merrell Dow Pharmaceuticals*, 509 US 579
Commonwealth of Puerto Rico et al. v. *The SS Zoe Colocotroni et al.* (628 F.2d 652)
Defenders of Wildlife v. *Andrus* (1977) 428 F. Supp.167
Duke Power v. *Carolina Environmental Study Group* (1978) 438 US 59
Elam v. *Alcolac Inc.* (1989) 765 SW 2d 42
Environmental Defence Fund v. *Masey* (1993) 986 F.2d 528, D.C. Cir.
Ethyl Corporation v. *EPA* (1976) F.2d 1, D.C. Cir.
Everett v. *Paschall* (1910) 61 Wash. 47, 111 P. 879
Frye v. *United States* (1923) 293 F.1013, D.C. Cir.
Georgia v. *Tennessee Copper Co.* [1907] 206 US 230, 27 S.Ct. 618
Greenpeace USA v. *Stone*, 748 F. Supp. 749
Guo Chun Di v. *Carroll* (1994) 842 F. Supp. 858
Hale v. *Henkel* (1906) 201 US 43
Hanly v. *Mitchell*, 409 US 990, 93 S.Ct. 313

Hodel v. *Virginia Surface Mining and Reclamation Association, Inc.* (1981) 452 US 264
In The Matter Of: Oil Spill By The Amoco Cadiz Off The Coast Of France On March 16, 1978, 954 F.2d 1279
Industrial Union Department, AFL-CIO v. *American Petroleum Institute* (1980) 448 US 607
International Harvester Co. Ruckelshaus (1973) 478 F.2d 615
Katz v. *United States*, 389 US 347
Kleppe v. *Sierra Club* (1976) 427 US 390
Lucas v. *South Carolina Coastal Council* (1992) 112 US 286
Lujan v. *Defenders of Wildlife*, 504 US 555; 112 S.Ct. 2130
Lujan v. *National Wildlife Federation* (1990) 110 S.Ct. 3177
Metropolitan Edison Co. v. *People Against Nuclear Energy* (1983) 460 US 766
NRDC v. *United States EPA*, 824 F.2d 1146
National Audubon Society v. *Hester*, 801 F.2d 405
National Audubon Society v. *Superior Court of Alpine County* (1983) 658 P.2d 709
Natural Resources Defense Council (NRDC), Inc. v. *Morton* (1972) 458 F.2d 827, 838 (D.C.Cir.)
New York v. *Shore Realty* (1985) 759 F.2d 1032, 1043–1044, 2d Cir.
Northern Spotted Owl Decision (1988) 716 F.Supp. 479
Organised Fishermen of Florida v. *Andrus* (1980) 488 F.Supp. 1351
Penn Central Transportation Co. v. *City of New York* (1978) 438 US 104
Pierson v. *Post*, 3 Cal. R. 175
Rector, Wardens and Vestry of St. Christopher's Episcopal Church v. *C.S. McCrossan, Inc.* (1975) 235 NW 2d 609, Minn.
Reserve Mining Co. v. *Environmental Protection Agency*, 514 F.2d 492
Robertson, Chief of United States Forest Service v. *Seattle Audubon Society* (1992) 503 US 429; 112 S.Ct. 1407
Robertson v. *Methow Valley Citizens Council* (1989) 109 S.Ct. 1835
Sierra Club v. *Morton* (1972) 405 US 727
Sierra Club v. *Froehlke* (1976) 534 F.2d 1289, 8th Cir.
Sindell v. *Abbott Laboratories* (1980) 163 Cal Rptr 132
Spur Industries v. *Del Webb Development Co.* 108 Ariz. 178, 494 P.2d 700
Stevens v. *City of Cannon Beach* (1994) 505 US 1207
Swift v. *Gifford* (1872) 23 F.Cas. 558 Mass.
Tennessee Valley Authority v. *Hill* (1978) 437 US 153, 98 S.Ct. 2279
Union Electric Co. v. *EPA*, 427 US 246
United States v. *Carolawn Chemical Company* (1988) No. C-88-94-L, D.N.H.
United States v. *Carroll Towing Co.* (1947) 159 F.2d 169 (2d Cir.)
United States v. *Fleet Factors* (1990) 901 F.2d 1550, 11th Cir.
United States v. *Maryland Bank & Trust* (1986) 632 F.Supp. 573, D. Md.
United States v. *Ottati and Goss* (1985) 630 F.Supp. 1360, 1395-96, D.N.H.
United States v. *Shell Oil* (1985) 605 F.Supp. 1064, 1073, D. Co.

United States v. *Students Challenging Agency Procedures* (1973) 412 U.S. 669
Whalen v. *Union Bag and Paper Co.* (1913) 101 NE 805
Whitney v. *United States* (1991) 926 F.2d 1169

Other cases

Akonaay and Lohay (1994) Court of Appeal of Tanzania at Dar es Salaam, Civil Appeal No. 31 of 1994
Advisory Opinion on the Legality of the Threat or Use of Nuclear Weapons (1996) ICJ Rep.
Australia v. *France*, *New Zealand* v. *France* (1974) ICJ Rep. 253
Australian Conservation Foundation v. *Commonwealth* (1980) 146 CLR 493
Case concerning the Gabcikovo-Nagymaros Project (1997) ICJ Rep.
Commonwealth of Puerto Rico et al. v. *The SS Zoe Colocotroni et al.* 628 F.2d 652
Corfu Channel Case (1949) ICJ Rep. 22
Dispute Panel Report on US Restrictions on Imports of Tuna (1994) 33 ILM 839
Dr. Mohiuddin Farooque v. *Bangladesh and others* (1997) 17 BLD (AD), Vol. XVII, pp.1 to 33
EC Measures Concerning Meat and Meat Products (Hormones), (3 TTR 2d 523)
Fellows v. *Sarina and others* (1996) NSW Lexis 2780
Fredin v. *Sweden* (1991) 13 EHRR 784
Friends of Hinchinbrook Society Inc v. *Minister For Environment and Others* (1997, 142 ALR 632)
Greenpeace Australia v. *Redbank Power Company* (1994) 86 LGERA 143
Juan Antonio Oposa and others v. *Hon'ble Fulgencio S. Factoran and another* (1994) 33 ILM 173
Kent and Others v. *James Cavanagh, Minister of State for Works and Others*, 1 ACTR 43
Lac Lanoux Arbitration (1957) ILR 24
Leatch v. *Director-General National Parks & Wildlife Service and Shoalhaven City Council* (1993) Land and Environment Court of New South Wales, 1993 NSW Lexis 8229
McIntyre v. *Christchurch City Council* (1996) NZRMA 289
M.C. Mehta v. *Union of India & Others* [1988] 2 SCR 530
M.C. Mehta v. *Union of India* [1992] AIR SC 382
Milirrpum v. *Nabalco City Council* (1971) 17 FLR 141
Nicholls v. *Director National Parks and Wildlife Service* (1994) 84 LGERA 397
North Coast Environment Council Incorporated v. *Minister for Resources* (1994) 16 December 1994, unreported
North Sea Continental Shelf Cases (1969) ICJ Rep 3
Ontario Ltd. v. *Metropolitan Toronto and Region Conservation Authority* [1996] O.J. No. 1392, 62 A.C.W.S. 3d 591

R. v. *Crown Zellerbach* [1988] 1 SCR 401
R. v. *Enso Forest Products Ltd* (1993) 85 B.C.L.R. (2d) 249
Re Mohr and Great Barrier Reef Marine Park Authority and Marcum and Others, 53 ALD 635
Robinson v. *Western Australian Museum* (1977) 138 CLR 283
Shirley Primary School v. *Telecom Mobile Communications Ltd* [1999] NZRMA 66
Tasmanian Conservation Trust Inc. v. *Minister For Resources and Gunns Ltd.* (1995) 37 ALD 73
The Anglo-Norwegian Fisheries Case (1951) ICJ Rep 116
The Commonwealth v. *The State of Tasmania* (1983) 158 CLR 1
The Commonwealth of Australia v. *The State of Tasmania*, 46 ALR 625
The Kaimanawa Wild Horse Preservation Society Inc. v. *Attorney-General* [1997] NZRMA 356
The Queen v. *Land Use Planning Review Panel ex Parte M F Cas Pty Ltd* (1998), unreported
The Trail Smelter Arbitration [1941] AJIL 684
UK v. *Albania* (1949) ICJ Rep 4
United States – Restrictions on Imports of Tuna (1991) 33 ILM 1598
Vellore Citizens' Welfare Foundation (1996) AIR SC 2715
Wi Parata v. *Bishop of Wellington* (1877) 3 NZ Jur (NS) SC 72

NZRMA	New Zealand Resource Management Appeals	New Zealand
P.2d	Pacific Reporter, Second Series	US
P&CR	Property and Compensation Reports	UK
QB	Law Reports: Queen's Bench Division	UK
S.C.	Scots Session Cases	UK
SCLR	Scottish Civil Law Reports	UK
SCR	Canadian Supreme Court Reports	Canada
S.Ct.	Supreme Court of the US	US
S.J.	Solicitors Journal	UK
SLT	Scots Law Times	UK
SW	Southwestern Reporter	US
TTR	Trading Law and Trading Law Reports	International
US	United States: Supreme Court Reports	US
WLR	Weekly Law Reports	UK
WR	Weekly Reporter	UK

Introduction

The aim of this book is to introduce environmental issues, and law, in an interdisciplinary context. It is primarily a text for readers who are interested in environmental protection but who have no prior legal background. As such, it seeks to provide a platform of information and analysis that will enable the reader to engage with more 'mainstream' environmental law sources.

The environment presents a particular and peculiar challenge to law and legal systems. The holistic view of environmental systems that emerged in the late twentieth century instructs us that *all* human activity impacts on 'nature' and the living world. Furthermore, as Barry Commoner (1971) reminds us, 'every separate entity is connected to all the rest'. The environment is not one of the traditional compartmentalised problems that law and legal systems are used to dealing with. Using law to solve environmental problems is not much like legislating for shop opening hours or the maximum speed of vehicles. A broad perspective must always be taken and yet focused and practical solutions must be generated. The environment poses lawmakers with a remarkable project.

A brief history of environment and law

Widespread public appreciation of environmental problems is fairly new: often dated to the publication of Rachel Carson's *Silent Spring* (1962), which warned of the dangers of the bioaccumulation and biomagnification of pesticides in the food web. Nevertheless, laws regulating the environment are of some antiquity. Since the earliest civilisations mankind has understood that protection of nature's basic commodities (land, water, soil, plants and animals) is necessary for long-term survival. Thus, every known civilisation has had rules prohibiting, to greater or lesser extent, unsustainable agriculture and husbandry practices. In ancient Babylon, over 4,000 years ago, the Code of Hammurabi sought to prevent environmental degradation through proper control of land, crops and (especially) irrigation (Johns 1999; Ware 1905). The Old Testament contains prescriptions for sustainable agriculture and distributional justice:

> For six years you are to sow your fields and harvest the crops, but during the seventh year let the land lie unplowed and unused. Then the poor among your people may get food from it, and the wild animals may eat what they leave. Do the same with your vineyard and your olive grove.
>
> (Exodus 23:10)

In Europe, Roman law evolved a highly developed system of water controls and laws governing matters such as rivers, sewers, drainage and reservoirs (Ware 1905). Many of these rules remain today, in modified form, in the codified laws of continental European countries.

England has a long history of concern for environmental issues (Clapp 1994; Thomas 1984). The Common Law, which spread throughout the land after the invasion of William of Normandy in 1066, established an elaborate system for protection of soil, air, water, crops and animals (Baker 1990). In time it established, through the action of 'nuisance', a prototype right to a clean environment (Cornish and Clark 1989, 154–8) although, due to the power relations and economic disparities existing between employer and employee, this was often effectively unusable by the ordinary citizen (Brenner 1974; McLaren 1983).

British anti-pollution legislation can be traced back to the 1388 Act for Punishing Nuisances which Cause Corruption of the Air near Cities and Great Towns that imposed fines and a clean-up obligation on anyone who 'cast and put in ditches, rivers and other waters',

> so much dung and other filth of the garbage and entrails as well as of beasts killed . . . that the air there is greatly corrupt and infect, and many maladies and other intolerable diseases do daily happen . . . to the inhabitants, dwellers, repairers, and travellers aforesaid.

The other major source of British anti-pollution legislation is the Victorian sanitary reform movement and its concomitant public health programmes. The cholera epidemics of 1831, 1854 and 1866 struck not only the poor but also the middle and upper classes (Underwood 1948); a lack of class immunity that stimulated research and legislative action.

Chadwick's sanitary reform movement and his seminal study, 'Report on the Sanitary Condition of the Labouring Population of Great Britain' (Chadwick 1842) provided the platform for a plethora of public cleanliness statutes including the Public Health Act 1848 and the Nuisances and Diseases Prevention Act 1848.

Britain was the first nation to have a fully functional environmental agency: the Alkali Inspectorate, formed in 1863 under the Alkali etc. Works Act. The Inspectorate's task was to administer the regulatory regime set up under that Act to reduce and, where possible, eliminate emissions of hydrochloric acid produced during the manufacture of sodium carbonate. The Inspectorate's responsibilities gradually grew as other substances and processes were added to its remit under further legislation. Eventually the Alkali Inspectorate was subsumed into what is now the Environment Agency in England and Wales, the Scottish Environmental Protection Agency in Scotland, and the Environment Department in Northern Ireland. In reality little significant improvement in air pollution occurred until the passage of the Clean Air Act 1956, passed in response to lethal smogs of the preceding year. It is possible to argue that the history of improvements in British air quality is largely a history of changes in technologies, fuels and processes: law mirrored this process but added only a minor forcing effect (Clapp 1994).

British conservation legislation has a substantial history, beginning with medieval royal regimes for management of forest game. The notion of a 'forest' was, in the late medieval period, not focused on extensive trees but, rather, implied extensive tracts of land in which royal hunting privileges, and correlative hunting laws, amounted to a primitive but effective conservation regime (Rackham 1976). Modern conservation laws can be traced back to the passage of the Wild Birds Protection Act 1904, which gave an unusually clear sign of the conservationist approaches that were to follow. Conservation efforts grew significantly after the close of the Second World War, with the National Parks and Access to the Countryside Act 1949 providing the essential structure (parks, nature reserves and sites of special scientific interest) that, although modified, remains to this day.

In the United States laws for environmental protection can be traced back at least as far as the late 1700s when, as a price of agreement to the Union, smaller states required large-scale transfers of land by the larger states to central government. These transfers enabled the creation, in time, of public lands which, in turn, allowed for the creation of the US system national parks (Futrell 1993). Another early element of US environmental law was concern with air pollution and, more specifically, smoke. Originally, both in the UK and in the US, smoke had been seen as a sign of economic growth and prosperity: the more the better, since it symbolised the economic status of a town or city. This perception was not, however, to last. By the late 1800s it had been widely recognised that the smoke and foul air in cities such as Pittsburgh, Chicago, St Louis and Cincinnati were not only aesthetic evils but also the major causes of respiratory disease in those areas. A change in national attitude towards smoke came about as the positive image was replaced by one in which smoke was ugly and (because of its negative health effects) uneconomic. Smoke began to cast a pall over the very nature of the Anglo-European project of civilisation through economic growth. Learning from the British experience, US political societies such as the Anti-Smoke League successfully pressed for smoke control legislation. Unfortunately these early efforts were largely ineffective because either the infrastructure or political will to enforce the legislation concerned did not exist. In the US (unlike the UK) judges were able to, and often did, declare such legislation unconstitutional (Stradling 1999). Proper control of US air pollution would wait until the passage of the Clean Air Act of 1970.

Early US conservation policies – hence laws – were strongly influenced by the American ideal of and relationship with nature. As Nash (1967) notes, 'wilderness was the basic ingredient of American civilization. From the raw materials of the physical wilderness Americans built a civilization; with the idea or symbol of wilderness they sought to give that civilization identity and meaning.' Wilderness was not, however, an object of reverence: indeed, early settlers put great efforts into the subjugation of the forests and wild animals (Taylor 1998).

Modern US environmental law can be traced back to the publication of Rachel Carson's influential book, *Silent Spring* (1962), warning of the hitherto ignored dangers of DDT and other pesticides. Her book's wider message was that human society was as sick and poisoned as the creatures to which it applied biocides. It was a society bound by an archaic and simplistic notion of the 'control of nature', belonging to a Neanderthal age when nature was supposed to exist solely for the purpose of man. Carson's ideas were the direct antecedents of the legal principles adopted by congress in the National Environmental Policy Act of 1969. Other significant formative influences in the history of US environmental law including: Hardin's 'Tragedy of the Commons' thesis, Kenneth

Boulding's notion of 'Spaceship Earth' (Boulding 1971), Christopher Stone's call for the rights of trees and other environmental entities, the rise of citizen rights to enforcement of laws.

The National Environmental Policy Act of 1969 was followed, rapidly, by a plethora of environmental legislation including the Clean Air Act, the Clean Water Act, the Coastal Zone Management Act, the Endangered Species Act, the Safe Drinking Water Act, the Resource Conservation and Recovery Act, the Toxic Substances Control Act, and the Comprehensive Environmental Response, Compensation and Liability Act. The United States' legislative programme for environmental protection is now amongst the most comprehensive of any country.

Complementing this diverse range of legal controls is an equally broad set of executive agencies with responsibilities for the administration of US environmental law (see Table 0.1). Of these the Environmental Protection Agency (EPA) is arguably the most important. This body differs from its UK counterpart by virtue of its much broader mandate to actually set legal standards, and issue regulations giving legal effect to these. The trend in the UK, in recent years, has been towards increasing unification of administration in keeping with the 'one-stop-shop' view of regulation.

The importance of environmental law

Let us consider why the environment needs law or, in other words, why environmental law is important. The need for strong and effective laws in the maintenance of a sustainable society has long been a theme of political and philosophical writing. In his great work, *The Republic*, Plato wrote of the need for effective lawmakers to prevent the decay of society into decadence and destitution. Hobbes, in *Leviathan* (1996, orig. 1651), took a similar view: that men, unless they are subject to a sovereign, will act to the detriment of one another so that their lives become 'solitary, poor, nasty, brutish, and short'. Whether we agree with this essentially pessimistic view of human nature or see humanity as essentially good and perfectible (e.g. Rogers 1961) it seems that humans are selfish enough to act in ways that lead to significant deterioration of the earth's environment, unless their actions are placed under legal limitation.

The idea that humanity, left to its own devices, will eventually (and inevitably) bring about environmental degradation has been expressed vividly in Garret Hardin's seminal essay 'The Tragedy of the Commons'

the Tragedy is 'mutual coercion, mutually agreed upon'. The term 'coercion' is, here, used broadly to include *legal regulation* of the commons, as well as the use of *economic instruments* such as taxes and charges (e.g., in Hardin's example, a 'cow tax').

Beyond coercion lie two radical solutions. One is to scrap the commons; that is, to divide the resource into several parcels and attach to each a bundle of *property rights*. Each herdsman, knowing the limited carrying capacity of his plot, would not act to his own long-term detriment by adding more and more cows. The reverse angle is to replace the private gain of herding with a form of *communal benefit sharing*. This finds expression in political and philosophical doctrines such as communism: 'From each according to his ability, to each according to his needs' (Marx 1875, in Tucker 1972). In this solution not only the dis-benefit but also the profits of additional cattle are shared amongst the whole community. Since utility and dis-utility accrue to a person in roughly equal proportions, the incentive to overload the system is, in principle, negated.

Each of these themes is discussed in greater detail later in this book: environmental education (Chapter 5); environmental ethics (Chapter 8); regulation, self-regulation and property rights (Chapter 5); economic instruments (Chapter 6); science (Chapter 7) and the failings of communal solutions (Chapter 9).

The major thesis of this book is that environmental law provides the overarching framework for the multidisciplinary approaches that are necessary if we are to halt the trend of environmental decay. Each strategy, on its own, cannot deal with all of the problems; together they can provide a comprehensive and holistic response. Although sociologists have suggested that ethics, economics, and education can more or less displace legal controls, the truth is that each strategy requires 'legal embeddedness' to have the necessary strength to succeed. The environment needs good law if it is to avoid suffering further serious harm.

Further Reading

Buck Cox, J. (1985) 'No Tragedy on the Commons', *Environmental Ethics*, vol. 7, pp. 49–61. Was Hardin right?

Futrell, J.W. (1993) 'The History of Environmental Law', in C. Campbell-Mohn, B. Breen and J.W. Futrell, *Environmental Law from Resources to Recovery*

(St Paul, Minn., West Publishing), and Brenner, J.F. (1974) 'Nuisance Law and the Industrial Revolution', *Journal of Legal Studies*, vol. 3, 403. These sources provide some interesting detail on historical aspects of legislative and judge-made environmental laws in the US and UK respectively.

1 Law and legal systems

- The concept of law
- Legal systems
- Divisions of legal power

Introduction

The environment is clearly at risk from a variety of sources of harm, mostly of human origin. In order to begin to tackle this problem it is important that we develop strategies for modifying human behaviour towards environmentally benign practices and away from environmentally damaging ones. In very broad terms, techniques for modifying human behaviour can be thought of as falling into two types: incentives (carrots) and disincentives (sticks). Law is important since it creates frameworks within which incentives and disincentives can operate. Quite commonly law is itself an incentive or disincentive. In other cases it operates as a structure in which to position non-legal techniques such as economic instruments or informational and educational measures.

Law is all-pervasive. Other methods for influencing human behaviour are, to a certain extent, voluntary or optional. Education, ethics, peer and family pressure: these all apply in varying degrees. Law, on the other hand, cannot easily be avoided. It is axiomatic to the 'rule of law' that law in a society applies equally to everyone at all times (Hutchinson and Monahan 1987). As such it can be a powerful weapon in the armoury against environmental degradation.

In this chapter we introduce a number of concepts of law and a variety of legal systems. The purpose is to explain what is meant by 'law' and how the types of legal system encountered can help or hinder the overall goal of environmental protection.

The concept of law

The study of the environment and the law immediately poses the question, 'What is law?' Answering this question is, in one sense, straightforward. Law has been described as 'generally . . . a way of regulating human behaviour' (McEldowney and McEldowney 1996, 3). Yet such simple formulations leave many issues unresolved. Is an edict from a government department a 'law'? Is policy guidance issued by a government department (e.g. Britain's Department of the Environment, Food and Rural Affairs) to be regarded as 'law'? Is the 'Precautionary Principle' part of the law? It is important to be able to distinguish between 'law' on the one hand, and other normative systems on the other. If we confuse non-legal and legal standards we fail to understand the best means of influencing the content and enforcement of each. Furthermore, laws often carry economic or penal penalties for transgression that non-laws do not. So, let us consider more closely the concept of 'law'.

Law as commands

One school of thought (Hobbes 1996, orig. 1651; Bentham 1891, orig. 1776; Austin 1954, orig. 1832) is that the only things that count as 'laws' are commands of a sovereign, backed up by sanctions in the event of disobedience. A sovereign, for Austin, is an individual or body that is clearly identifiable, habitually obeyed by society, and is not habitually obedient to any other superior. In the UK, Parliament would be such an institution.

One problem with the command concept of law is that it doesn't fit very readily with laws that merely empower or permit one to do something. The right to freedom of speech as, for instance, found in the First Amendment to the United States Constitution, is not a *command.* Similarly, when environmentalists speak of a 'right of access to environmental information' they are not implying that they are, in any sense, required to access such data. The command concept of law has another difficulty: it fails adequately to separate *legal* coercion from non-legal coercion. A mugger – who is habitually obeyed but who obeys no other person – might stop a member of the public and demand money on pain of death: is this *law*? A multinational corporation might instruct its staff to keep quiet about pollution incidents caused by its activities, on pain of dismissal. Would that be *law*?

Law as rules

Problems with 'command' theories of law led to the development of 'rule' theories of law. Hart (1961), the most eminent rule theorist, divided legal rules into *primary* rules and *secondary* rules. Primary rules have substantive content (e.g. it is an offence to pollute a watercourse). Secondary rules are rules about primary rules. They tell us how to recognise, introduce, alter, or remove primary rules. An example of a secondary rule – in this case the *rule of recognition* – is that in the UK a valid statute requires passage of a bill through both Houses of Parliament and receipt of the Royal Assent. It is the possession of both primary and secondary rules which, according to Hart, demarcates a *legal system* from other institutions for social control. This implies, incidentally, that less formal systems of social conventions and rules such as those possessed by certain indigenous peoples may not achieve the status of 'legal system'.

The rule model of law faces certain problems. First, what should courts do if the law does not contain a rule governing a particular case or if the rule seems vague? Hart's answer is that laws, whilst generally comprehensive and clear, will nevertheless entail a 'penumbra of doubt' where judges must exercise *discretion*. This would imply that we must accept that judges actually make law where the legislature has been unclear or left a gap. The discretion explanation itself, however, is subject to criticism. In both the UK and the United States judges in most of the higher courts are appointed rather than elected. It may be asked why judges, who lack democratic legitimacy, should be able to exercise discretion so as to create law.

Second, it is not certain that any clear rules exist. Some rules are made not by the legislature but by judges. In the case of judge-made rules (known as 'precedents') the scope of any given rule is often unclear. A rule which, on its face, appears to be narrow in its application, may subsequently turn out to be broadly interpreted. A revealing instance is the case of *Donoghue* v. *Stevenson* (1932). Here a woman consumed ginger beer bought for her by her friend, containing a decomposing snail. The snail's remains caused the woman to fall ill and she successfully claimed damages against the drink manufacturer. Prior to this case the House of Lords had not generally allowed damage claims where the injured party lacked 'privity of contract' with the seller. Thus the decision in *Donoghue* had the effect of creating a new rule of liability. At the time of the judgment it appeared that only a narrow rule had been created (i.e. that beverage manufacturers have a duty to ensure that their drinks are safe for consumption). In fact the case later turned out to have created a very broad rule indeed. *Donoghue* v.

Stevenson was the genesis of the whole body of law that today is know as *negligence* – now the largest area of civil litigation – which extends a duty of care to a vast range of actors and activities.

Similar arguments may be made in the case of rules made by the legislature (i.e. legislation). It is difficult to predict the meaning that courts will attribute to words used in legislation. So-called 'realist' writers go so far as to deny that law contains any rules as such, as distinct from the actual operations of the courts which are only ever imperfectly predictable (Frank 1970).

Law as principles

Not everyone agrees that law consists simply of a body of clear rules surrounded by a woolly mantle of judicial discretion. Dworkin (1977), for one, famously argued that law also contains *principles* and does not contain discretion.

Dworkin distinguished rules and principles as follows. He said that rules apply in an 'all or nothing' fashion (e.g. river pollution is forbidden) whereas principles have the quality of 'weight'; that is to say, a principle is never absolute and is always subject to being balanced with and against other principles. An example of a principle might be that 'a person shall not profit from his or her own wrong' (Dworkin 1977, 22) or 'a polluter shall pay for environmental damage caused'.

Unlike Hart, Dworkin denied that judges have discretion when faced with unclear or seemingly unjust cases. Instead he asserted that, in such hard cases, judges should reach a solution based on the *principles* of their particular legal system.

The sorts of principles which can be found in most legal systems include:

- the principle of *proportionality* (laws should not be Draconian);
- the principle of *non-discrimination* (laws should not make arbitrary distinctions between subjects);
- principles of natural justice (e.g. the right to a fair trial); and
- *equitable principles* such as 'equity will not permit a statute to be used as an instrument of fraud'.

An important question that we examine in Chapter 4 is whether *environmental* principles such as the Precautionary Principle and the Polluter Pays Principle have also been accepted into environmental law.

Laws based on local custom – 'customary law' – continue to be of considerable practical importance in many developing countries, especially in Africa. Individuals often rely on customary rights to protect their environment, and their own homes, from the threat of development. A useful example is *Akonaay and Lohay* (1994). The Tanzanian Regulation of Land Tenure (Established Villages) Act 1992 declared customary rights in land to be extinct, and prohibited the payment of compensation for such extinction. Akonaay and Lohay rejected this usurpation of their legal rights and sought to recover land that had been taken from them under colonial rule. They based their claim on the fact that they had improved the land, and were by that reason alone entitled to compensation. Furthermore, under the terms of the Tanzanian constitution, they could not be deprived of land without compensation. The Attorney-General attempted to rebut their claim, contending that customary rights of occupancy were not 'property rights' and thus were not protected by the constitution. The Tanzanian Court of Appeal, influenced by the academic Allen (1993), found the 1992 law to be unconstitutional. Customary rights of occupancy were 'property': to have determined otherwise would have led to the absurd conclusion that the vast majority of Tanzanians were no more than squatters in their own country. Additionally, the court added, property should be recognised where a person has worked on the land to improve it and, thereby, add value to it. The court, in this last respect, unconsciously reflected the writings of the English philosopher John Locke (1988, orig. 1690).

Many important concepts existing within one legal culture may be absent, or present only in altered form, in others. In particular, indigenous people often consider land and the flora and fauna that inhabit land to contain the spirits of ancestors, and that the land belongs truly to ancestors and to generations as yet unborn (David and Brierley 1985, 550). Such conceptions of nature tend to be minimally exploitative and tend to further sustainable modes of living. Aboriginal Australians, for instance, traditionally had no concept of private property rights in land – at least rights that are exclusionary between members of each tribal group and which are freely transferable, as in the western concept of property (Dodson 1994). This resulted in the non-recognition of Aboriginal customary rights of access to grazing lands (*Milirrpum* v. *Nabalco City Council* (1971)) (see Gray 1994, 182).

The relationship between law and traditions varies between societies. In the US, for example, resort to law and litigation has long been commonplace. America is a notoriously litigious society. In Japan, by

Box 1.1

Indigenous Maoris and environmental theft

The legal marginalisation of indigenous people, and the suppression of their right to the environment, is nowhere clearer than in Maori history. The 1840 Treaty of Waitangi, signed between Governor Hobson and 500 Maori chiefs, was never intended to give the English Crown full sovereignty over the islands (Williams and Grinlinton 1997). More plausibly, the Treaty intended only to cede governorship ('Kawanatanga') in return for the guarantee, in article II of the Treaty, that the Maori people would retain chieftainship ('Rangatiratanga') over their land, resources and spiritual treasures. The New Zealand courts, however, regarded the Treaty as of no legal force and allowed successive New Zealand governments to grant land rights to New Zealand citizens, thus confiscating Maori lands of great ecological and spiritual significance. Notably the Maori Affairs Amendment Act 1967 compulsorily 'converted' more than 96,000 hectares of Maori land to non-Maori ownership (Kirkpatrick 1997). Furthermore, abrogated Maori property rights could not be subject to litigation (*Wi Parata* v. *Bishop of Wellington* (1877)), and the Native Land Act 1909 statutorily barred claims that native land rights had been improperly extinguished.

Fortunately, the tide turned with the setting up of the Waitangi Tribunal, under the Treaty of Waitangi Act 1975, to redress wrongs done to the Maori people in contravention of their Treaty rights. More recently, New Zealand courts have begun to recognise Maori rights of possession and use (Williams and Grinlinton 1997, 21). Future alienation of Maori land is now stringently restricted under the Maori Land Act 1993. Other Acts now give Maori customs and law much higher priority. For instance, the New Zealand Resource Management Act 1991 – a superbly holistic piece of environmental legislation – recognises the Treaty of Waitangi and gives the Maori a right to be consulted in any decisions made under the Act. The Act also requires all persons exercising its functions and powers to 'recognise' and 'provide for', amongst other things, 'the relationship of Maori and their culture and traditions with their ancestral lands, water, sites, waahi tapu, and other taonga' (section 6). Mining on the Maori indigenous land is now limited by the Crown Minerals Act 1991 which allows the Maori people to indicate lands of particular cultural importance that should be excluded from the mineral extraction programmes.

contrast, tradition, conciliation and mediation are the preferred conflict resolution mechanisms. Sometimes law cannot replace the social functions of tradition and custom. It is notable that victory in the compensation cases that followed the mercury poisoning incident at Minimata Bay did not satisfy the victims' feelings of anger. These privately issued reports and summaries. In the UK environmental cases are reported either in full or in summary in *Environmental Law Reports*, *Environmental Law Review*, the *Journal of Environmental Law*, the *Journal of Planning and Environment Law* and *Environmental Law and Management*. Some of these are available in full text format on the Internet, through services such as NESLI (the National Electronic Site Licence Initiative, http://www.nesli.ac.uk/cgi-bin/loginpage).

In addition to the traditional medium of the printed page, decided cases are increasingly disseminated via electronic media such as CD roms and the Internet. Many academic institutions, especially university law faculties, have access to Internet-based resources such as:

JUSTIS	www.justis.com/database/case_law.html
LEXIS	www.lexis.com
Westlaw	www.westlaw.com and www.westlaw.co.uk

LEXIS and Westlaw are powerful comprehensive electronic databases of decided cases (Heesterman 1993; LEXIS 1999). Whereas traditional paper-based reports tend to include only those cases that, in the view of the series editors, involve important points of law, LEXIS and Westlaw include all decided cases which meet certain criteria. They also allow the user to interrogate the database using keywords combined in logical format.

Law distinguished from policy

An important distinction in the concept of law question is whether, and if so how, laws can be distinguished from policies. It is now a settled principle that European Union (EU) laws – 'Directives' – cannot be given effect simply by administrative means. Government circulars, strategies or advice documents cannot substitute for the hard-edged character of legislation which is necessary so that 'individuals are in a position to know their rights in order to rely upon them where appropriate' (*Commission* v. *Germany* [1991]). Two factors distinguish law from policy. First, policy is generally *advisory* in nature, recommending objectives or setting targets, rather than prescribing particular actions. Second, policy may derive from any number of institutional processes whereas law, as we have already mentioned (see 'Law as rules', pp. 12–13) must pass strict secondary rules of recognition before it has legal quality.

The 'relegation' of some instrument to the field of policy rather than law does not exclude it from legal importance. Failure to take relevant policies into account or, conversely, consideration of irrelevant policies, may invalidate decisions of public bodies (*Anisminic Ltd* v. *Foreign Compensation Commission* [1969]). Not surprisingly, disputes not infrequently arise concerning the relevance, hence permissibility, of environmental policies taken into account by public authorities. In *The Queen* v. *Land Use Planning Review Panel ex Parte M F Cas Pty Ltd* (1998), for instance, the applicants applied for, but were refused, permission to re-zone an area of Tasmanian bushland to allow development. In reaching their decision the Planning Review Panel took the Precautionary Principle into account. The applicants argued that by so doing the Panel had made reference to an irrelevant consideration. The Supreme Court rejected that claim: the panel had simply erred on the side of caution and had acted in accordance with 'good common sense'.

Sometimes environmental policies *must* be taken into account. For instance, in UK development control law, government advice about development controls, issued in the form of Planning Policy Guidance (PPG) notes, must be taken into consideration in the determination of applications for planning permission (Moore 1987, 176).

Legal systems

Laws for environmental protection are embedded in, and are influenced by the characteristics of, 'parent' legal systems. A law that exists for the purpose of environmental improvement can only function well if the legal system in which it exists is strong and effective. It is, therefore, useful to examine legal systems in general and to try to point out the manner in which, as systems, they influence the matter of environmental protection.

Traditionally, legal systems have been classified along two axes: first, *geographical scope*; second, *internal characteristics*. Geographically, legal systems exist at several levels:

1 Local or sub-regional (e.g. state, province or *Länder*).
2 National (e.g. UK).
3 Regional (e.g. EU).
4 International.

A detailed discussion of international law is postponed until Chapter 4. In what follows, the discussion is limited to local, national and regional legal systems.

Two factors strongly influence the internal characteristics of a legal system. The first is the relationship between the parts and the whole; that is, their *relational construction* (composite or unitary systems). The second is the dominant scheme *within* a given state for law-making and application; that is, the *genre* of the legal system (e.g. common law, civil law). These factors are now considered in more depth, taking the genre first and the relational construction second.

Federal legal systems

National legal systems differ markedly according to their division into *unitary* or *composite* constitutions. In unitary constitutions laws are made by one political entity. An example of a unitary constitution is Ireland in which all laws are made by the Oireachtas. The Oireachtas consists of the President and two Houses, Dáil Éireann (House of Representatives) and Seanad Éireann (Senate).

Composite legal systems, by contrast, have more than one level of political authority and, correspondingly, more than one set of law-making institutions. Within composite systems, two dominant models are encountered: *federal* systems and *devolved* systems. Both involve sharing of political authority between the centre and sub-regions of a nation, but they differ in certain attributes.

Federal systems divide power, on a permanent or semi-permanent basis, between two sovereigns, or supreme lawmakers: central (federal) government and the local (state) government. This allows the creation of a nation whilst at the same time preserving cultural and regional differences, and facilitating the participation of geographic minorities in the affairs of government (Sampford 1990).

Federalism's *decentralised* law-making function has several virtues for the environment. It can accommodate differing environmental values – that is, promote diversity (Stewart 1977) – and give local populations a say in environmental affairs. It can allow regions to engage in what Brandeis called 'social and economic experiments' (Friendly 1977) which, in the context of the environment, might include trials of different levels of environmental integrity. By locating governance nearer to smaller population units it is, in principle, in keeping with broadly anarchistic visions of a sustainable society (Schumacher 1973) – although, of course, not everyone agrees that 'small is beautiful' (Beckerman 1995).

Federalism's *centralised* law-making function allows for economies of scale. Centralised solutions are often highly desirable responses to the transnational nature of environmental problems. Pollution knows no arbitrary or administrative boundaries and must often, therefore, be dealt with in a harmonised manner. Stewart (1977) argues that centralised law-making permits society to engage in 'the politics of sacrifice'. Local areas will, he argues, be more willing to make material sacrifices to protect the environment for non-instrumental ethical reasons (e.g. because of a perceived duty to future generations) when they know that all other local areas are going to be called upon to make the same contribution. Furthermore, federal politicians, being relatively remote and less sensitive to local opposition, will be better placed and more willing to pass strong environmental laws that may affect local populations more than state politicians.

Some federal systems – e.g. Australia (Rothwell and Boer 1998) and Germany (Winter 1994) – share power in environmental law-making fairly evenly between the federal and state governments. Sometimes, as in India, this is done by creating lists of central, provincial and joint competence (Anderson 1998). Other systems simply limit the legislative power of central government to those situations in which it has been given an explicit mandate to act. This is the case in both the United States and the European Union.

In the United States, Amendment X to the Constitution reserves all powers not delegated to the United States by the Constitution 'to the States respectively or to the people'. Despite this apparent bias towards state legislation, in practice a broad federal competence for environmental protection has developed from the 'commerce clause' (Article I, section 8) of the US Constitution. This provides that

> The Congress shall have Power . . . To regulate Commerce with foreign Nations, and among the several States, and with the Indian Tribes . . .

Since most environmental regulations impact directly or indirectly on inter-state commerce, the Supreme Court rarely determines environmental laws to be beyond federal competence. *Hodel* v. *Virginia Surface Mining and Reclamation Association, Inc* (1981) is illustrative. In this case a mining company challenged the constitutionality of the Surface Mining Control and Reclamation Act of 1977 – a congressional statute regulating strip mining – on the grounds that it amounted to regulation of the use of private land. Accordingly, the mining company maintained, the matter should fall within state police powers, not federal control. The Supreme Court, however, took the view that in interpreting the commerce clause it was bound to defer to the pre-existing congressional finding that mining has economic, social and environmental impacts affecting commerce between states.

This tendency towards central government control over most laws for environmental protection is also to be observed in Canadian environmental law. As the Canadian Supreme Court now acknowledges (*R* v. *Crown Zellerbach* [1988]), over time issues such as transregional pollution can 'migrate' from provincial to federal jurisdiction. The government in Ottawa adds to this tendency by exerting fiscal and political pressures to bring provincial environmental policy in line with federal objectives: so-called 'co-operative federalism' (Kibel 1999, 77).

In the European Union (EU) – a federal institution in all but name – the appropriate level for governance is determined, in theory at least, by application of the doctrine of *subsidiarity* (Adonis 1991; Emilou 1992). Subsidiarity is a presumption in favour of law-creation *at the level nearest to the people* (i.e. by EU Member States themselves, rather than by EU institutions). Subsidiarity and its 'twin' principle of proportionality are concretised in Article 5 (formerly 3b) of the EU Treaty:

> The Community shall act within the limits of the powers conferred upon it by this Treaty and of the objectives assigned to it therein. In areas which do not fall within its exclusive competence, the Community shall take action, in accordance with the principle of subsidiarity, only if and in so far as the objectives of the proposed action cannot be sufficiently achieved by the Member States and can therefore, by reason of the scale or effects of the proposed action, be better achieved by the Community.
>
> Any action by the Community shall not go beyond what is necessary to achieve the objectives of this Treaty.

The reference to 'limits of the powers' is a reminder that, when the EU creates law, it must base such law on the correct provision of the Community Treaty. Any EU law that is not can be struck down as lacking a proper legal base (e.g. Case C-376/98 – no proper legal basis for Directive restricting tobacco advertising).

Unfortunately, no guidance is given as to the meaning of the term 'better achieved' in Article 5, and so the concept is open to competing interpretations. Axelrod (1994, 117) comments: '"Better" is ultimately a political choice and a matter of subjective interpretation.' This implies that the European Court of Justice would be unlikely to entertain claims that, in enacting environmental laws, the EU had overstepped the mark.

Certain commentators have tended to downplay the potential effect of subsidiarity on the grounds that there are few areas of environmental law in which the Community is not better placed than Member States to achieve objectives (Wilkinson 1992). Brinkhorst (1993), for one, argues that EU environmental law should generally be formed at the regional level because (a) most conservation or pollution effects are transboundary in nature and (b) allowing action by individual states would upset the 'level playing field' of economic costs. There is little point protecting migratory birds in one Member State if they are shot whilst migrating through another, or in prohibiting harmful emissions from incinerators if the territory concerned is affected by drifting fumes from across the border. In practice the European Commission tends to simply ignore the 'better than' test when proposing new environmental legislation, instead justifying new law (merely) on the grounds that 'all Member States are concerned by this action' (Somsen and Sprokkereef 1996, 47).

Subsidiarity has some positive environmental effects. It allows 'greener' Member States (e.g. Germany) to exceed the harmonised EU environmental standards (as permitted under Article 176 EC Treaty). It also warrants a larger role for local areas *within* individual states, thus fitting the green epithet 'think globally, act locally' (Collier 1997; Golub 1996). On the other hand subsidiarity has, on occasions, been used to block environmental progress. In 1992 the UK proposed scrapping twenty-seven items of EU environmental law on the grounds that these related to matters better dealt with at Member State level (Wils 1994). Certain aspects of animal welfare with no transboundary impact – e.g. bullfighting – have been characterised as not meriting legal harmonisation at European level (Brinkhorst 1993). A proposed Directive on minimum standards for zoos was downgraded to a mere 'recommendation', a proposal for an EU carbon tax was rejected (Collier 1997, 10), and the EU Commission has decided not to deal with 'local' issues such as promoting bicycle use and reducing disco noise (Krämer 1995).

Is federalism, overall, of advantage to the environment? In addition to the theoretical advantages discussed above, Bodansky and Brunnée (1998) observe that federal legislation has, in some instances, been used to legislate to protect natural resources against state development as, for example, in *The Commonwealth of Australia* v. *The State of Tasmania*. Despite these benefits, the answer to the question is not clear. One problem is that federal laws may prioritise free trade over environmental protection. Illustrating this trend is the case *City of Philadelphia* v. *New York* (1978) which determined that New Jersey could not prohibit waste importation from other states.

Another difficulty is that federal systems can lead to disintegrated solutions in which some but not all sub-regions produce environmental programmes. A similar charge is made against the 'cooperative' federalism of the Australian federal Environment Protection and Biodiversity Conservation Act 1999. Hughes (1999) concludes that because the 1999 Act lacks a *national* environmental impact assessment system it is unlikely to provide the level of protection which would be expected for a major overhaul of Commonwealth environmental law.

A related difficulty is that of delay. Considerable time can pass between the creation of an obligation by the nation-state, and the implementation of the laws necessary to give effect to that obligation at the lower sub-national level. Layers of bureaucracy may have to be passed through as the instructions 'filter down'. The sub-regions may have to ratify or approve the measure. Belgium, for example, suffers from this problem and has been brought before the ECJ on many occasions due to its inability to provide timely implementation of EU environmental Directives (CEC 1990).

The position afforded to the environment in federal systems depends significantly on the attitude taken by the Supreme Court on the interpretation of legislation, and the question of constitutionality. A striking example of an empowering decision is *The Commonwealth* v. *The State of Tasmania* (1983) in which the Australian High Court upheld the constitutionality of the World Heritage Act which implements the UNESCO Convention Concerning the Protection of the World Cultural and Natural Heritage.

The US Supreme Court, after passing through a period of pro-environment attitude in the 1970s, has gradually moved to a more pro-business stance. Indicative of the earlier attitude is *Tennessee Valley Authority* v. *Hill* (1978). Work on the Tellico Dam – a federally funded regional development project – was halted due to the discovery of a small fish, the 'snail darter', in the Little Tennessee River where the dam was being constructed. The Secretary of the Interior declared the snail darter to be an endangered species; as such, the fish was protected by the Endangered Species Act 1973. A majority of the Supreme Court upheld a decision to halt work on the project in the light of this discovery. The majority reasoned it made no difference how close to completion the project was. Congress, in passing the 1973 Act, had been fully aware that the legislation would produce results 'requiring the sacrifice of the anticipated benefits of the project and of many millions of dollars in public funds'.

The pro-environment stance evidenced in the US Supreme Court and many lower federal courts was, in no small measure, influenced by Judge Harold Leventhal's 'Hardlook' school of judicial review of agency decisions. The Hardlook philosophy is the notion that, in reviewing environmental decisions made by public agencies, the courts should not interject their own views. Rather, they should ensure that the agency has not just noticed, but given serious consideration to the contentions of citizens, litigants, and the environmental consequences of the decision in hand (see e.g. *Natural Resources Defense Council (NRDC), Inc.* v. *Morton* (1972) and *Kleppe* v. *Sierra Club* (1976).

Representative of the trend away from rigorous protection of environmental values is the Supreme Court decision in *Robertson, Chief of United States Forest Service* v. *Seattle Audubon Society* (1992). This case arose from protests by environmentalists over the harvesting of old-growth forests in the Pacific Northwest. Environmentalists protested that this habitat removal would result in the indirect deaths of spotted owls. Accordingly, they challenged the logging as illegal under several existing environmental statutes: the Migratory Bird Treaty Act, the National Environmental Policy Act and the National Forest Management Act. However, before the litigation could be brought to a conclusion, Congress effectively 'pulled the plug' by enacting Section 318 of the Department of the Interior and Related Agencies Appropriations Act 1990, otherwise known as the 'Northwest Timber Compromise'. The 'compromise' entailed by s. 318 was that timber harvesting and sales would be required to go ahead in 'exchange' for total protection of relatively small areas of owl habitat. Controversially, Section 318 (b)(6)(A) of the 1990 Act stated that management of forests according to the Compromise was 'adequate consideration for the purpose of meeting the statutory requirements that are the basis for the [legal challenges mounted by the Audubon Society and others]'. The Audubon Society viewed this as interference, by Congress, in the outcome of legal proceedings – as tantamount to legislating for the outcome of a case. If this claim was true the 'compromise' would certainly have been unconstitutional. The Supreme Court decided that it did not direct the outcome of litigation, but in fact merely made an alteration to the law that underlay the ongoing litigation.

As such the new law was constitutional and the Audubon Society was unable to continue its claims that timber harvesting conflicted with the protective requirements of existing environmental laws.

Devolved legal systems

In contrast to federal systems, *devolved* legal systems presume reservation of political authority to central government, with sub-regions having only such authority as is specifically delegated to them from time to time. Thus, the legislative power of the sub-regions is always precarious and subject to withdrawal or amendment.

The UK is a good example of a devolved composite legal system (see Olowofoyeku 1999). The United Kingdom of Great Britain and Northern Ireland has evolved since early times and comprises England, Scotland, Wales and Northern Ireland. The devolution of each is briefly considered below.

Scotland and England were constitutionally united by the Act of Union of 1707, which abolished the Scottish Parliament in Edinburgh and gave Scotland representation at the Westminster Parliament in London. Scotland, however, retained its own legal system, which differs from that of England and Wales in many areas of law (e.g. differences in rules for civil liability for environmental damage). Separation was also effected by, in many cases, enactment of separate Scottish legislation. The existence of this distinct body of law administered by independent Scottish regulators – Scottish Environmental Protection Agency (pollution controls) and Scottish National Heritage (conservation controls) – has led to a distinctly Scottish regime for environmental protection (Reid [1992] 1997; Smith *et al.* 1997).

Devolution for Scotland came about through the Scotland Act 1998. This created the Scottish Parliament: a body that is competent to legislate on any matter unless certain exceptions apply. These include that the Scottish legislation in question relates to ‘reserved matters’. Reserved matters include: transport of radioactive materials, pollution associated with oil and gas exploration, ownership and exploitation of coal, and external trade in endangered species (Schedule 5). Scottish judges are empowered to strike down Acts of the Scottish Parliament as unconstitutional if they go beyond the powers of the devolution legislation or if they conflict with the European Convention on Human Rights. As the latter is growing in importance as a source of environmental law (see Chapter 3) this is the first step towards a constitutional guarantee of environmental rights.

Clearly, Scotland is now in a position to create its own environmental laws, including matters such as town and country planning, pollution controls, animal protection and movements, flood prevention and conservation. The question is how it will exercise this power and whether the result will further differentiate Scottish environmental protection from that which applies in the rest of the UK. Research into Scottish attitudes to the environment reveal a complex pattern of values which differ in some respects to those held in the rest of Britain (McCormick and McDowell 1999). So far, however, the environment has not been high on the Scottish legislative agenda (ENDS 1999a). Predictions for future developments include Poustie's (1999) anticipation of 'more distinctively Scottish policies and guidance', Brown (1998) foresees 'great scope for a particularly Scottish approach to the implementation into national law of future EU legislation on civil liability for environmental harm'. Little (2000), on the other hand, considers that the Scottish Executive and Parliament will not find it easy to flex their environmental muscles due to the cooling effect of the potential for political review by the Westminster Government and the terms of the Memorandum and Concordats that accompanied devolution.

Wales was annexed by force in the thirteenth century by Edward I and, from the death of Llewelyn in 1282, ceased to have any separate political existence. The Act of Union of 1536 extended English law to the principality and constitutionally united Wales to England. Through this union, statutes passed by the Westminster Parliament govern Welsh affairs. Welsh devolution was achieved through the Government of Wales Act 1998 that established a National Assembly for Wales. Section 21 of the 1998 Act gives the Assembly limited functions (reform of Welsh health authorities; reform of other Welsh public bodies; support of culture, etc. and consideration of matters affecting Wales), but other functions can be transferred by an Order of the Queen in Council under Section 22. Over 5,000 functions given to the Secretary of State for Wales under almost 350 Acts of Parliament, as well as under European Union legislation, have been transferred in this way. Section 121 of the 1998 Act requires the Assembly to produce a scheme for promoting sustainable development.

The new arrangements are also likely to increase Welsh local authorities' roles in setting and implementing local environment policies (Pontin 1999). However, as with Scotland, it seems that, so far, environmental matters are not of pressing concern to the Welsh Assembly (ENDS 1999a). It is notable that the form and extent of Welsh devolution is much restricted in comparison to that afforded to Scotland: there is no power, for instance, to enact primary Welsh legislation. Consequently, it is likely that environmental law in England and Wales will, for the foreseeable future, be fairly homogeneous.

Northern Ireland is the product of a long complex Anglo-Irish history (Buckland 1981). The English conquest of Ireland, begun in the twelfth century, was complete by the end of the sixteenth. Civil uprisings led to a scheme for partition government and this, in turn, was followed by the passage, in Westminster, of the Government of Ireland Act 1920. The 1920 Act set up two subordinate Parliaments in Ireland: one for six counties (also known as 'Northern Ireland'), another for twenty-six counties of what was, until 1948, the Free State (also known as the 'Republic of Ireland'). The six counties remained under direct British rule with about 80 per cent of the government powers reserved to Westminster, the rest being devolved to the Northern Ireland Parliament, 'Stormont'. Political failures, as well as increasing violence, led to the re-imposition of direct rule through the Northern Ireland (Temporary Provisions) Act 1972. Northern Ireland legislation, after the re-imposition of direct rule, is made under the Northern Ireland Act 1974. These orders, which usually mirror British law, require approval by the UK Parliament. Environmental laws, as well as other laws for the Province, are invariably delayed by this process. Paradoxically, Northern Ireland's environment has remained in relatively good shape because of the lack of economic development resulting from the troubles in the Province (Morrow and Turner 1998).

The long struggle in Northern Ireland between nationalist and pro-UK forces led, eventually, to the Good Friday Agreement and the Northern Ireland Act 1998. The 1998 Act set up the Northern Ireland Assembly with full legislative competence except on certain reserved matters (e.g. international relations, defence of the realm, control of nuclear, biological and chemical weapons and other weapons of mass destruction, dignities, treason, and immigration). It remains to be seen whether this body will legislate for better environmental protection.

The overall effect of the devolution process in the UK may be that, as Wilson (1999, 12) suggests, 'the reordering of many priorities in the environmental field' will result in a much less homogeneous end result – one much more akin to the federal system that obtains in the United States.

Common law legal systems

At national level legal systems can be divided into at least three genres (David and Brierley 1985):

(a) common law systems;
(b) civil law (or Romano-German) systems; and
(c) religious legal systems

Until the fall of the Soviet Union and the communist regimes in the states of Eastern Europe, it was also common to refer to a fourth class: socialist legal systems. These, however, have largely disappeared; they are, therefore, not discussed here. Aspects of the legal systems of the remainder socialist states – Cuba, China and North Korea – which affect the environment, are discussed in Chapter 9.

The term 'common law' refers to a legal system in which one body of law is common to all parts of a given territory. Common law originated in England where, prior to the Norman Conquest, what were in effect several kingdoms operated separate legal systems based on local customs and traditions. After the Conquest the law of the Royal Courts became common throughout the land (Baker 1990). Common law countries include England and Wales (Scotland is a hybrid common law–civil law system), Australia, Nigeria, Kenya, Zambia, the United States of America, New Zealand, Canada, and various Pacific and East Asian countries.

Common law systems are characterised by the prevalence of judge-made law (i.e. *the doctrine of legal precedent*). Under the doctrine of precedent, judges – although reluctant to admit the fact – make new law through the determination of legal disputes ('cases'). These cases constitute rules that must be followed by all courts of equal or lower standing in the hierarchy of courts. Even though environmental protection is now largely a body of statutory and administrative law, the existence or absence of judge-made rules is still important.

The simplified diagram (Figure 1.1) indicates the hierarchy of criminal courts in the UK and their functions.

Referring to Figure 1.1, rules made by judges in the House of Lords bind the judiciary in the Court of Appeal and the High Court; rules made in the latter bind the Crown Court and the Magistrates' Courts.

Theoretically, the doctrine of precedent must be strictly applied, but in practice the courts often find ways around it. In some cases, subsequent

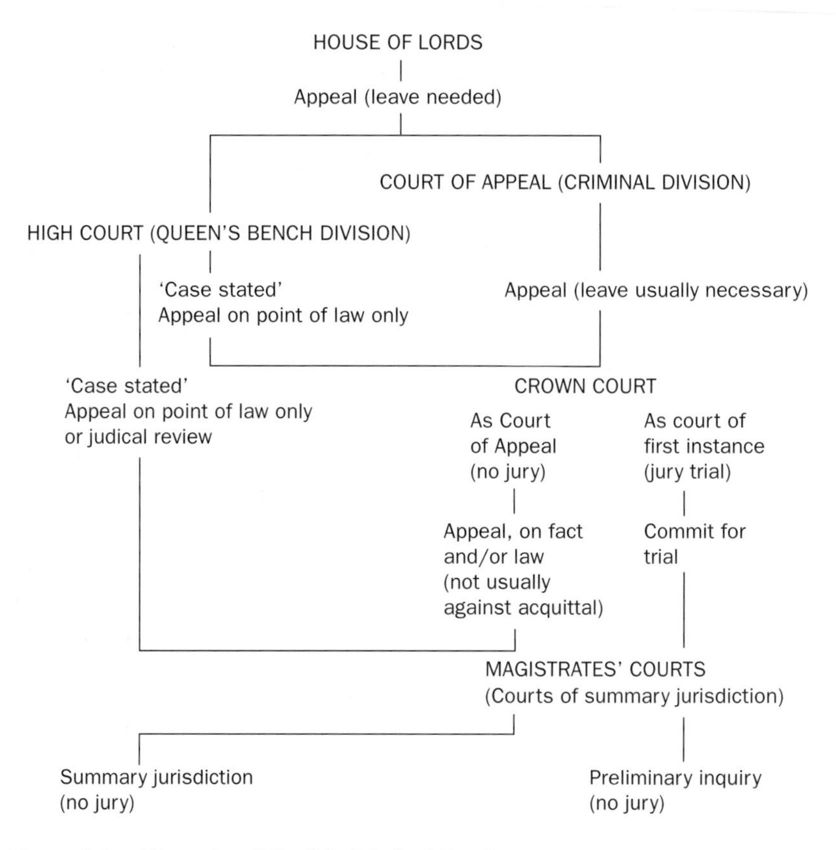

Figure 1.1 ***Hierarchy of English Criminal Courts***

judgments interpret the applicability of the precedent in question very narrowly. Another evasion mechanism is the technique known as 'distinguishing the case'. This occurs when judges decide that the facts of the present case are so materially different from those of the precedent that the precedent need not apply. A further 'escape mechanism' is the later creation of exceptions to the rule established by the precedent. The creation of exceptions to precedents, and limits and exceptions to the exceptions, vastly increases the complexity of judge-made law.

Consistent with the emphasis on judge-made law, common law nations have historically favoured case law over legislation. Today, however, this bias is reduced – regrettably on some accounts (Calabresi 1991). In part, this is due to the complexity of the subject matter – commonly, statutes create detailed bodies of rules often containing over a hundred sections and several thousand words. In part, it is due to the increasing reluctance of judges in some jurisdictions to play an active role in the creation of new legal principles. English judges, for instance, have become rather timid.

In some countries judges have found the new area of environmental law to be one in which they feel able to exert considerable influence. Robinson (1995) contends that:

> Judges can play the crucial role of helping to *establish*, not merely act upon society's values with regard to the environment. They can do this by giving legal cladding to environmental principles such as sustainability, intergenerational equity and . . . the precautionary principle.
>
> (Robinson 1995, 314)

India is a good example. India is 'blessed with judiciary willing to introduce innovative and daring judicial procedures so as to protect the environment' (Lau 1995). Indian Supreme Court judges felt able, during the 1970s, to create rules extending eligibility for legal aid and substantially modifying the rules of 'standing' (i.e. the prerequisites to bringing a legal action). These steps allowed the poor and oppressed to petition the court (Dias 1994). In Australia judges have creatively utilised international environmental law as a guide to statutory interpretation and as a mechanism for reviewing domestic legislation (Rothwell and Boer 1988).

A second main characteristic of common law systems is that court structures tend to be relatively undifferentiated. For example, criminal cases of all kinds, including environmental crimes, are heard in the UK criminal courts illustrated in Figure 1.1. In such a non-differentiated system judges, especially those in the lower level courts, may hear too few environmental cases to build up real expertise in environmental matters.

Three principal solutions exist to the above concern. The first, and least radical, would be to compose a list of specifically 'environmental' judges who would sit in all environmental cases (McLeod 1995). A more radical remedy would be to create an environmental division of the existing High Court (Carnwarth 1992). Finally, a full UK environmental court could be instituted, along the lines of the New South Wales Land and Environment Court created in 1979. The NSW court has wide powers to hear pollution, planning and conservation cases and has generally been considered to be a success (Steiner 1995).

A UK environment court could be a flexible body, able to draw on expertise from other professions (architects, surveyors, etc.) and appoint 'multi-disciplinary adjudicating panels' (Woolf 1992). It could integrate scientific experts to help deal with complex scientific and technical issues arising in environmental cases (McAuslan 1991). The recent report by Grant (1999) for the UK's Department of Environment Transport and the Regions points out that creation of an environment court might address two other concerns. First, it could resolve the apparent failure, by the UK, to comply with Article 6 of the European Convention on Human Rights, which protects the right to a fair trial:

> in the determination of his civil rights and obligations or of any criminal charge against him, everyone is entitled to a fair and public hearing within a reasonable time by an independent and impartial tribunal established by law . . .

Currently, many planning disputes are dealt with by the Planning Inspectorate. Grant (1999) doubts whether this body is sufficiently independent from Government, and whether a Planning Inspectorate inquiry is a 'fair and public hearing' sufficient to comply with Article 6.

Second, the UK is a signatory to the Aarhus Convention on public access to justice. This creates two key obligations: first, to ensure access to a review procedure before a court of law and/or another independent and impartial body established by law; second, to allow access by the public to administrative and/or judicial procedures in order to challenge acts or omissions which contravene provisions of national law relating to the environment. An environment court could provide much fuller compliance with these duties.

On the other hand Bates (1993), writing against situating criminal cases in an environmental court, maintains that the rigours and complexities of criminal environmental law are better dealt with in the mainstream legal system. More recently the Select Committee on Environment, Transport and Regional Affairs (1999), has warned that such a body might, if not properly designed, create problems. Unless carefully linked to the existing Planning Inspectorate it might be seen as diminishing the role of local authorities in the planning process as well as worsening existing perceptions of an over-legalistic planning appeal process.

Civil law legal systems

Civil law legal systems (named after the Roman law applicable between citizens: *jus civile*) are characterised by systematic 'codification' (i.e.

writing down of their ordinary law), rather than (as in Common Law countries) the creation of law through accumulated decisions of judges. Civil law systems include countries which have drawn strongly on their Roman legal heritage in addition to other sources, and countries where Roman influence was weaker but whose law is principally created by legislation (for example, Scandinavian countries).

The civil law family of states is very large. Major civil law states include: Argentina, Austria, Belgium, Brazil, Bulgaria, Central African Republic, Chile, Colombia, Costa Rica, Croatia, Czech Republic, Denmark, Dominican Republic, Ecuador, El Salvador, Finland, France, Germany, Greece, Guatemala, Hungary, South Korea, Latvia, Luxembourg, Macedonia, Madagascar, Mexico, Moldova, Monaco, Netherlands, Nicaragua, Norway, Panama, Paraguay, Peru, Poland, Portugal, Romania, Russia, Serbia and Monténégro, Slovakia, Slovenia, Spain, Suriname, Sweden, Switzerland, Turkey, Ukraine, Uruguay, Venezuela and Vietnam.

Civil law systems have several distinguishing characteristics. First, law is contained in large written texts or 'codes'. These codes, rather than previous case law, mark a break with the past and act as a starting point for future legal interpretation. In German law, for example, most of the general law is found in the German Civil Code (BGB), which is abstract and complex (hence difficult for the layman to interpret) but also very precise.

The existence of basic codes does not preclude further and subsequent legislation. Indeed, being a relatively new area of law, environmental law is often contained in specialised legislation, which exists alongside the main codes. In Germany, in the period 1980–1993, over twenty environmental statutes were created (Winter 1994) including the Water Management Act (Wasserhaushaltsgesetz, WHG), the Drinking Water Act (Trinkwasserverordnung), the Federal Nature Conservation Act (Bundesnaturschutzgesetz), the Federal Pollution Protection Act (Bundesimmissionsschutzgesetz, BimSchG), the Waste Disposal Act (Abfallsbeseitigungsgesetz, AbfG), the Federal Building Act (Bundesbaugesetz, BbauG), the Packaging Act (Verpackungsverordnung, VerpackV) and the 1990 Environmental Liability Act (Umwelthaftungsgesetz). The existence of a more or less comprehensive legislative framework for environmental protection means that few areas are left unregulated, hence unprotected. The large number of overlapping statutes derives from the dual approach of German law: protecting the *media* (air, land and water) and concurrently regulating environmentally hazardous *processes* (Winter 1994, 1).

A second distinguishing feature of civil law systems is that the judiciary in such systems perceive their role to be more limited than has traditionally been the case in common law systems. Thus, civil law judges are not as openly 'activist' as their common law counterparts. To illustrate: German judges have refused to expand citizen's rights to challenge environmentally damaging decisions except in cases in which the citizen can show infringement of some private right or protective norm (Greve 1989).

Often, it is said that civil law judges perceive the codes and legislation rather than their own previous pronouncements to be the basis of the law. Civil law judges interpret articles of the basic codes and subsequent legislation by reference to the underlying intention of the legislature and not the literal meaning of words (the so-called *teleological approach*) (de Cruz 1995). In practice, however, the priority of the code breaks down because of the need for judicial interpretation. This, inevitably, reintroduces judicial creativity. In Germany, for instance, previous decisions are usually considered and judges attempt to achieve consistency – a weak but distinguishable form of the 'doctrine of precedent' (see pp. 31–2) (Foster 1993).

Civil law systems often take a different perspective on the correct structure of environmental law. Civil law systems usually distinguish between private law (e.g. contract) and public law (e.g. administrative law and criminal law) and have different courts for each 'stream' of law. These distinctions of substance and process have the potential to create problems when environmental protection is at stake. German courts, for instance, have sometimes subordinated criminal law to administrative law with the consequence that environmental crimes can be excused by reference to defective or illegally granted permits (Paeffgen 1991).

Religious legal systems

Some states have legal systems that share common features with their religious doctrines. This is especially true of Muslim, Hindu, and Buddhist countries in which law, as well as religious life, is more or less based on holy teachings. In Islamic states, for example, the teaching of the Islamic prophet Mohammed as revealed in the Qu'rān (Coulson 1978; Ajijola 1989) is the basis of law.

States which operate a Muslim or Islamic legal system, either in total or mixed with some other system, include Afghanistan, Algeria, Bahrain,

Bangladesh, Brunei, Egypt, Eritrea, Ethiopia, Gambia, Indonesia, Iran, Iraq, Lebanon, Mauritania, Morocco, Niger, Oman, Pakistan, Saudi Arabia, Somalia, Sudan, Syria, Tunisia, United Arab Emirates, and The Yemen.

Environmental protection in Islamic states is seen as an obligation of each individual, as part of the man's status as Caliph or steward for Allah (Dutton 1996). Legal judgments and rules are a part of *fiqh* – the putting of religious beliefs into action. Islamic law divides land into two categories *āmir* ('developed') and *mawāt* ('undeveloped'). *Āmir* may be the private property of the individual, but *mawāt* includes wilderness which cannot be owned. Water, pasture and firewood are available to all where they occur on undeveloped land. Hunting wild animals is permitted by the *sharī'a* so long as those animals killed for food are slaughtered quickly and efficiently, using a sharp knife, mentioning the name of Allah, and not in front of other animals (Dutton 1996).

From the Islamic perspective the environmental crisis is a result of human inaction or action. Islam contains a rule against usury or *ribā* (i.e. earning money purely from interest or money lending). Dutton (1996) observes that environmental damage caused by overdevelopment can be attributed to *ribā* since much development that takes place occurs only because it is actively promoted by banks and the international monetary system in order to stimulate demand for loans, from which they earn money. He contends that Islam offers an escape from this cycle of need-creation and degradation.

Conclusion

We have seen that, from theoretical perspectives, there are differing views about the concept of law itself. Commands, rules and principles all feature in an integrated notion of law. The advantage of this mixed approach is that it allows varying degrees of coercion and judicial freedom in the upholding of environmental standards. The disadvantage of western conceptions of law is the inadvertent marginalisation of the laws and codes of indigenous peoples, whose life practices are so often much more sustainable than our own.

The type of legal system within which a given environmental law originates is of considerable importance in determining the ability of that law to tackle the environmental problem it is designed for. Unitary,

2 Legal concepts of environment

- **Differing concepts of 'the environment'**
- **The kinds of entities that may be of legal relevance**
- **Extraterrestrial environments**
- **Local environments versus one global environment**

Introduction

This chapter explores the concept of the environment, its legal treatment and manifestation. Whether law is successful in protection of the environment will depend significantly upon the range of entities that it is able to protect. Law can only succeed in protecting an object if that object is clearly defined in law. This is exemplified in the well-known law, present in all legal systems, against murder. Murder, it is generally accepted, is the unlawful killing of *a person* accompanied by a certain mental state (roughly speaking 'intention'). In order for this to be effective there has to be clarity about the meaning of 'person'. Is, for instance, an unborn child covered by this term? If so, how can abortion be legally justified? Are persons in permanent comas or other forms of severe brain function depletion to be regarded as legal persons? By analogy it is apparent that any body of law which seeks to protect the environment needs to have a workable and sufficiently broad conception of the subject of its concern.

Generally, it may be best to adopt a concept of environment that is broad in scope. If one defines 'environment' too narrowly then it may not be possible for law to make a meaningful contribution to the resolution of ecological degradation. This chapter therefore encourages critical thinking about the legal concept of 'the environment' and about a broad range of entities that *could* be included in such a concept.

The concept of the environment

Literally, 'environment' means 'that which surrounds' so, in a sense, the environment is the whole physical universe: Einstein reportedly said 'the environment is everything that is not me', although even that formulation overlooks micro-organisms living symbiotically or parasitically *within* our bodies (Botkin 1990, 136; Lovelock 1979, 102). In addition to the physical universe, there is another major class of things: *abstract* entities (e.g. truth, love, redness, society). It is important to note that abstract and physical entities may co-exist (e.g. a red flower).

The two classes are amenable to further differentiation into *living* and *non-living* forms. Living things are characterised by three qualities: (a) use of energy; (b) organisation of material or information; (c) ability to replicate (Silver 1999). This definition requires us to accept as alive some odd things such as viruses and, perhaps one day, self-replicating robots.

Table 2.1 ***A taxonomy of the environment***

Possible components	*Physical*	*Abstract*
Non-living	e.g. a rock or heat	e.g. truth
Living	e.g. a horse	e.g. an ecosystem

Table 2.1 illustrates the combination of the above classificatory scheme of entities.

Physical non-living components

Land

That land, air and water are indisputably components of the environment is reflected in Section 1 of the UK's Environmental Protection Act 1990. This declares that:

> [t]he 'environment' consists of all, or any, of the following media, namely, the air, water and land; and the medium of air includes the air within buildings and the air within other natural or man-made structures above or below ground.

'Land' is a generic term including rock, soil, sand, etc. as well as, more generally, the things attached to it. Of these soil is most often singled out for detailed legal protection – see, for example, the 1981 United Nations World Soil Charter; Chapters 10 to 14 of Agenda 21 of the 1992 United

Nations Conference on Environment and Development; and the European Soil Charter. The latter declares that

> Soil is one of humanity's most precious assets. It allows plants, animals and man to live on the earth's surface. Soil is a living and dynamic medium which supports plant and animal life. It is vital to man's existence as a source of goods and raw materials. It is a fundamental part of the biosphere and, together with vegetation and climate, helps to regulate the circulation and affects the quality of water. Soil is an entity in itself. As it contains traces of the evolution of the earth and its living creatures, and is the basic element of the landscape, its scientific and cultural investment must be taken into consideration.

Despite such insightful declarations few comprehensive soil laws exist. Generally, soil only becomes a subject of legal concern when it is contaminated with toxic substances, when it is used for growing sensitive crops (e.g. EU Regulation 2092/91 on Organic Production of Agricultural Products), or when it faces erosion risks (e.g. the 1994 United Nations Convention to Combat Desertification).

Underground strata are, in common law jurisdictions, dealt with through the legal maxim *cujus est solum, ejus est usque ad coelum et ad inferos* (the owner of the soil owns all up to the sky and down to the centre of the earth). In England and Wales, with certain exceptions (gold, silver, coal, oil), underground strata and minerals belong to the owner of the surface land. Underground 'space' (e.g. caverns or chambers) usually remains in the ownership of the person who owns the surface of the soil, but it can be transferred separately from the surface so that title lies in different hands (Gray 1999).

Land that lies under watercourses (i.e. river or lake beds) is an ecologically important feature of the environment: it provides habitat for water plants and animals, spawning grounds for fish and, by the build up of sediment, sequestrates pollutants. Legally, land under watercourses is often dealt with by extension of the rules dealing with dry land. In the English common law the bed of a non-tidal river is presumed to belong to the owner or owners of the riverbank on either side (*Blount* v. *Lanyard* [1891]). Interference with a riverbed is trespass and, as such, legally actionable by the owner of that land.

The 71 per cent of land that lies under the oceans might also be considered as part of a broad concept of the environment. In recent years laws have been developed controlling exploitation of the seabed.

Simplifying matters somewhat, these laws differentiate between the seabed of the territorial sea, and the seabed of the high seas.

The United Nations Convention on the Law of the Sea 1982 (UNCLOS) provides that, in the territorial sea (i.e. up to twelve nautical miles from shore) the coastal state has 'sovereign rights for the purpose of exploring and exploiting, conserving and managing the living natural resources of the sea bed and subsoil and superadjacent waters'. In English law this is reflected in the rule that the whole of the seabed within territorial limits belongs to the Crown, as do minerals lying under that seabed (*Lord Advocate* v. *Wemyss* [1900]). Thus all utilisation of the seabed, e.g. for anchoring fish farming cages, requires a Crown lease (*Re Shetland Salmon Farmers Association and Others* [1991]).

The deep-sea bed is governed by UNCLOS Part XI (WWF: 1994) as modified by the 1994 Agreement Relating to the Implementation of Part XI of the Convention on the Law of the Sea. Part XI declares the deep-sea bed to be 'the common heritage of mankind'. This implies that the benefits of its exploitation should be shared equitably amongst the nations. Part XI of UNCLOS and the 1994 Agreement create a detailed regime for the exploitation of nodules, found on the deep-sea bed, containing high-grade metals. These contain principally manganese, but also significant amounts of cobalt, copper, nickel and iron. Due to their depth, exploitation of such nodules is not likely to be economic until 2010 or later. When that state of technology is reached, UNCLOS Part XI will prevent exploitation without authorisation of the International Sea Bed Authority (ISBA). The ISBA will redistribute a share of the profits of exploitation among non-exploiting states in order to ensure that the rewards of exploitation do not accrue solely to those nations who happen to acquire the 'harvesting' technology first. Special consideration is to be given to land-based, metal-exporting states such as Zaire (cobalt), Gabon (manganese) and Canada (nickel) to offset the losses that those states will suffer if and when the demand for their exports drops as a result of seabed mining.

Notice that the definition of environment included in the UK Environmental Protection Act 1990, given above, includes land but does not expressly include things situated on that land such as buildings (although the air inside buildings is mentioned). Should such artefacts be thought of as part of the environment? As a matter of common law, things attached to land are regarded as *fixtures* (i.e. legally part of the parcel of land to which they are attached). This is reflected, in English law, in the

Box 2.1

Extraterrestrial environments?

In the future, progress in space travel, especially the possibility of lunar bases (see Lunar Corp 1999) may necessitate consideration of the status of 'the environment' on other celestial bodies. One question that looms for this era is whether such environments can be owned, either individually or by collective bodies such as states.

Historically there has been resistance to the idea of extraterrestrial property rights. The 1967 Treaty on Principles Governing the Activities of States in the Exploration and Use of Outer Space, including the Moon and Other Celestial Bodies, declares that 'Outer space, including the moon and other celestial bodies, is not subject to national appropriation by claim of sovereignty, by means of use or occupation, or by any other means' (Art. 2). More recently there has been a trend away from this absolute rejection of proprietary rights. One strand of this shift has been the argument that conceptualising the moon or other planets as 'common property' might have harmful environmental consequences. The drive to visit lunar and other celestial surfaces will almost certainly be economically led: quite possibly for mineral exploration or as a base for low-gravity science bases. The benefits of exploitation of the lunar landscape would accrue to individual lunar colonists whilst the negative effects would be shared by all present and many future colonists: the classic 'tragedy of the commons' situation which, in the absence of some other form of control, leads to overexploitation (Hardin 1968). Partly in response to that concern, and partly in order to create incentives for private space exploration, there is now a move towards privatisation of 'land' and property in space. It is already possible, in theory, to purchase property on the moon (Moonshop 1999), although perhaps only direct occupation of the lunar surface will suffice to create a true lunar property rights (Granqvist 1996). Roberts (1997) concludes that 'Enhancements are possible [to the legal regime governing solar-system environments] by implementing market-based standards for private property, explicitly permitting a private right of action for liability and creating a flexible procedure for the regulation of presently unknown environmental risks not addressed by other means.'

The prospect of lunar occupation raises the question of protection of the lunar environment. Article 7 of the Agreement Governing the Activities of States on the Moon and Other Celestial Bodies already requires that, 'In exploring and using the moon, State Parties shall take measures to prevent the disruption of the existing balance of its environment, whether by introducing adverse changes in that environment, by its harmful contamination through the introduction of extra-environmental matter or otherwise . . .'

The Treaty also has information requirements (Art. 7(2)), provisions for the creation of Areas of Special Scientific Interest (Art. 7(3)), as well as a general admonishment to pay due regard to the interests of future generations in connection with the exploration and use of the solar system (Art. 7(4)).

In the near future the concept of 'the environment' may have to be stretched to allow for simple forms of extraterrestrial life. Mars is a not unlikely spot for such life to turn up. One of the three life-detecting experiments from the first Viking Lander produced a positive result (Anonymous 1986). Fossil evidence from meteorites indicates the previous existence of Martian proto-bacteria (NASA: 1996).

In the longer term, contact with higher forms of extraterrestrial life *may* necessitate consideration of legal issues as (i) overlap of jurisdictions on celestial bodies, and (ii) 'aliens-rights'. Should extraterrestrial individuals be granted the same rights as humans? Piradov (1976, 114) thinks so. Granqvist (1996) suggests that this is premature since alien civilisations will perhaps need more or less rights than humans. Furthermore, humans may be the ones needing protection from aliens, not vice versa.

Law of Property Act 1925, Section 62, which automatically includes fixtures in a sale of land.

Environmental legislation sometimes expressly contrasts 'the environment', on the one hand, with buildings or structures, on the other. In the British Columbian case *R* v. *Enso Forest Products Ltd* (1993) the mill owner, who had allowed oil to escape into a ditch situated within the general boundary of his site, was charged with allowing a polluting substance to escape into 'the environment'. The Appeal Court decided that this was not an introduction of waste into 'the environment'. The relevant legislation contained mutually exclusive definitions of 'environment' and 'works' and the ditch, in the court's opinion, was properly characterised as the latter.

In favour of inclusion of buildings in the concept of the environment is the fact that, at least in western societies, buildings are the most significant 'surrounding' (with vehicles coming a close second). Ethical arguments have been made for the preservation of cities as a whole (Jamieson 1984), as well as of their culturally significant parts. Cities may also be indirectly valuable parts of the environment: being compact areas of high population density they preserve the relatively unbuilt, low-population character of surrounding rural areas.

Land and buildings are, along with other aspects of human-altered nature, the critical components of our notion of *landscape*. Very little unaltered nature remains (McKibben 1990) and often those aspects which have changed the least are most highly valued. Common law rules have done little to protect landscapes, since there can be no property in a spectacle (*Victoria Park Racing and Recreation Grounds Co Ltd* v. *Taylor* (1937)), no claim for interference with a view (Woodbury 1987), and no general prohibition on development (*St Helens Smelting* v. *Tipping* (1865)). Yet, ironically, it is those past developments and 'spoiled views' that the common law has permitted that now provide the landscapes we seek to protect. As Vogel has it, 'Nature is an historical entity' (1996, 139).

In recent times statute law has protected landscapes which are considered especially worthy of retention. In England, development control under the Town and Country Planning Act 1990 and the 'green belt' policy provides, with more or less success, containment of urban sprawl. Currently, only a minority of environmental laws explicitly include landscape in the definition of the environment. A rare example is Australia's Environment Protection and Biodiversity Conservation Act 1999, s. 528, which defines 'environment' to include 'the qualities and characteristics of locations, places and areas'. Of more direct importance is the recent Draft European Landscape Convention (http://www.coe.fr/cplre/eng/etxt/einstrjur/epaysage.htm) through which each Party agrees to 'ensure landscape protection, management and planning through the introduction of national measures and the organisation of European co-operation'.

Water

Clearly, any tolerably broad concept of the environment must include water in its several forms. We have already noted that water includes both marine and freshwater. But freshwater exists in forms other than rivers and lakes, e.g. as groundwater in aquifers. Groundwater is usually the focus of specific environmental laws, rather than a subject for inclusion in a general concept of the environment. The European Community Directive on the Protection of Groundwater Against Pollution Caused by Certain Dangerous Substances 80/68/EEC defines groundwater as 'all water which is below the surface of the ground in the saturation zone and in direct contact with the ground or subsoil', and prohibits the introduction of certain harmful substances into that water. In English common law a landowner may abstract unlimited quantities of groundwater (*Bradford Corporation* v. *Pickles* [1895]) and may sue in nuisance for the foreseeable pollution of such water (*Cambridge Water* v. *Eastern Counties Leather* [1994]). A landowner cannot, however, prevent the depleting effects of neighbouring landowners' groundwater abstractions (*Chasemore* v. *Richards* (1859)) (for the situation in the United States see Goldfarb 1988). Ownership of groundwater is, by virtue of this rule, only partial – a person only owns those molecules of water actually abstracted.

Water exists as atmospheric water vapour. Clouds play a vital role in maintaining global temperature as well as in their more obvious function as the origin of rain or other precipitation. Clouds are not yet, to my knowledge, included in legal concepts of environment. Interestingly, there are programmes by states to control the weather, typically through the use of cloud-seeding programmes (www.wmi.cban.com). These may eventually have to be brought under legal control since stimulating rain in one area may trigger drought in another (Taubenfield 1968; Rabie and Loubser 1990). The precipitation of atmospheric water vapour can be harmful if that vapour contains, either in solution or as particulate matter, harmful substances. Acid rain is a common problem in states that receive water vapour from other industrialised states and has been subject to legal controls (e.g. *The Trail Smelter Arbitration* [1941], and the 1979 Geneva Long-Range Transboundary Air Pollution Convention).

Finally, water can exist in the form of ice. Ice and snow form unique environments. They are the predominant element in high mountain regions, in high Arctic latitudes, and Antarctica. Each of these have their own legal frameworks: e.g.The Convention on the Regulation of Antarctic Mineral Resource Activities (establishing a moratorium on mineral exploitation in Antarctica – see Kimball 1993).

Air

Inclusion of air in the concept of the environment is non-controversial. There are, however, aspects of air that we may wish to consider more fully. The first is whether air inside buildings should be included. If we consider buildings to be part of the environment then there seems no reason not to do so. The UK Environmental Protection Act 1990, as we have seen, expressly includes air in buildings. The second aspect is the upper atmosphere. The gaseous composition of the upper atmosphere is so unlike that found at ground level that it stretches the term 'air' itself. Legislation designed to regulate the environment often does not indicate whether higher regions of the atmosphere are to be included. The upper atmosphere is particularly at risk from damage due to ozone-depleting substances and emissions from jet aircraft engines – a form of pollution which, to date, has received little legal attention.

How far up into the atmosphere does the notion of 'air' extend? When does it become 'space'? In legal literature, competing theories locate the boundary of atmosphere and outer space according to the maximum height of aircraft (i.e. about 85 km: Haley 1963) or the minimum height of rapidly moving satellites (i.e. about 100 km: Benkö *et al.* 1985). Of more practical significance is that, at low levels, the presence of aircraft has necessitated retraction from the *cujus est solum* maxim. In most countries ownership of land now includes airspace only as far as is necessary for the ordinary use and enjoyment of the land (*Baron Bernstein of Leigh* v. *Skyviews & General Ltd.* [1978] – aircraft taking photographs did not constitute trespassing; also Civil Aviation Act 1947, s. 40). Similarly, technology, not law, limits the height to which buildings may be constructed, even if this results in the overlooking of neighbouring land (Gray 1991, 255). The actual air itself within these regions is only *qualified* property. In particular, an owner of the land has a limited interest in the quality of such air as happens to present over that land at any moment in time – there is no absolute ownership of the molecules of air themselves (Gray 1991, 256).

Related to the inclusion of the atmosphere as a component of the environment is the idea of an *electromagnetic environment* (Singh 1987). This area, invisible to the human eye, consists of the electromagnetic fields surrounding the earth and the electromagnetic signals that are propagated through the normal physical media. Knowledge about the role of the earth's magnetic fields in maintaining the global ecosystem is currently fairly rudimentary. It is clear, however, that weather patterns and climatic changes are affected by shifts in the earth's magnetic field. Solar radiation from sun flares and sunspots contributes to climate change by blocking the formation of clouds which, in turn, results in the reflection of less solar heat back out into space. Certain bandwidths of electromagnetic radiation are harmful to humans and other animal life. Microwaves, for example, generated during the operation of mobile telephones, may be linked to brain tumours. There is also a growing body of scientific evidence, including a recent UK study (Ahuja 2001), suggesting that electromagnetic radiation from high voltage power lines is a factor in precipitating certain types of cancer (for further discussion of legal and scientific responses to electromagnetic radiation risks see Chapter 7).

An important subset of the electromagnetic environment is the *radiocommunications environment* (i.e. the portion of the electromagnetic spectrum in which radio waves can be propagated). This zone is regulated at national level and internationally through the International Telecommunications Union. The radiocommunications environment is highly dependent on placing satellites in the *geostationary orbit*: a narrow band around the earth at an altitude of about 36,000 km in which objects are stationary relative to the ground. If the geostationary orbit becomes full, or crowded, development of telecommunications via satellites may have to be limited or be subject to some kind of property or permits regime. The geostationary orbit is increasingly important for the placing of satellites for the surveillance of environmental changes. Satellite data may also have a role to play in the formation and enforcement of international environmental law, as well as in domestic litigation (Hodge 1997).

The current legal status of the geostationary zone is disputed. Most commentators consider it to be part of 'outer space', hence unowned, but in 1976 several equatorial states issued the 'Bogota Declaration' claiming property in the portion of the geostationary orbit that lies over their territory: a conflict that has yet to be resolved.

Physical living components

Any sensible concept of the environment will include not only the three environmental media – land, air and water – but also those organisms dependent upon them. The Albanian Law on Environmental Protection, 1991, Article 2, for instance, states that

> 'Environment' means all the natural and anthropic elements and factors, in their action and interaction,

and goes on to define 'natural elements' as

> water, air, soil and subsoil, solar radiation, vegetable and animal organisms, with all natural processes and phenomena generated by their interaction and which affect life . . .

Should the living elements of the environment include only *wild* plants and animals? This question is not easily answered. We may feel caught between the observation that nature in any pure form is virtually extinct (McKibben 1990), and ambivalence about whether domesticated organisms should be considered part of the environment. Weiner (1995) and Silver (1999) point out that domesticated plants and animals are all products of human manipulation of their genetic constituents. It would, therefore, be inconsistent to accept their inclusion in 'nature' whilst arguing inclusion of the products of biotechnology, such as genetically modified organisms.

There is one particular species of animal the inclusion of which in the concept of environment is problematic: *Homo sapiens sapiens*. Environmentalists, especially those of a humanist persuasion, are often keen to argue that the separation of humans and nature is the root of many of our current environmental problems (e.g. Naess 1989). But, if taken seriously, the re-incorporation of humans as 'just another species' may have challenging implications:

> Environmentalists often express regret that we human beings find it so hard to remember that we are part of nature – one species among many others – rather than something standing outside of nature . . . [but] *if we are part of nature, then everything we do is part of nature, and is natural in that primary sense*. When we domesticate organisms and bring them into a state of dependence on us, this is simply an example of one species exerting a selection pressure on another. If one calls this 'unnatural', one might just as well say the same of parasitism or symbiosis (compare human domestication of animals and plants and 'slave-making' in the social insects).
>
> (Sober 1995, 234)

Is this true? If so, perhaps pollution and species extinction are simply examples of Darwinian natural selection. Weiner (1995) argues strongly that the insights of the 'new ecology' (i.e. that humans are simply *part* of nature and that nature is fluid not static) leave us with many difficult questions unanswered, including why human intervention in the environment is wrong or should be legally proscribed.

Certain laws make it clear that humans are to be included in the environment concept. The Environmental Protection Act 1990 defines 'environmental harm' so as to include offence to human senses or human property (s. 29(4)). A closely related question is whether the environment should be considered part of a right to (human) life. In *G. and E.* v. *Norway* (Desgagné 1995) the European Court of Human Rights agreed that the human right to life, protected by the convention, includes a right to lead a nomadic existence in close contact with nature.

Abstract Components

Objective environment versus subjective environments

An important aspect of the philosophical debate concerning the concept of the environment is whether concepts such as 'environment', 'wilderness' and 'nature' have any objective basis, or whether they are social constructs (Rolston 1997; Vogel 1996). Cooper (1992) draws attention to a different sense of 'the environment' as 'a field of significance' – i.e. those aspects of place that an animal knows its way around and feels at home in. A field of significance is, therefore, always subjective (although it may also be *inter*-subjective). Cooper argues that undue emphasis on (a) the *physical* sense of environment and (b) a single *global* environment, fails to give adequate weight to concerns such as animal welfare and economic justice for the Third World. Cooper also thinks that it encourages an inappropriate idea of nature as a 'whole' object with which we need to develop a sense of oneness. Such oneness, he thinks, is not possible because we are separate from nature.

Dower (1994) cautions that the notion of a 'field of significance' should be tailored to the level of causal influence. This is not least because many people suffer from a mismatch between their fields of significance and the actual levels of causes of environmental detriment. Dower also points out that the two senses of 'environment' are interconnected: there could not be fields of significance without a supporting physical environment.

The value of a notion of a subjective, psychological environment is encountered in the United States' National Environmental Policy Act, requiring assessment of 'environmental impact' for federal projects that are 'significantly affecting the quality of the human environment'. The Supreme Court in *Hanly* v. *Mitchell* had to consider whether an impact statement should have been compiled in relation to the construction of a jail in the back of a nine-storey court house in Manhattan. The Court held that 'the Act must be construed to include protection of the *quality of life* for city residents' (emphasis added). In *Metropolitan Edison Co.* v. *People Against Nuclear Energy* (1983), addressing whether EIA was required prior to the re-opening of the Three Mile Island nuclear plant, the court agreed that psychological stresses and injuries are cognisable under NEPA although, on the facts, the plaintiffs were suffering only from the impact of the *perception* of risk.

Local environments versus one global environment

The Dower/Cooper contribution, above, links into a further issue: should we be concerned with (multiple) local environments or a (single) global environment? Preservation of local environments may be the only way to achieve real improvements in the overall environment – as in the green catchphrase 'think globally, act locally'. However, not all agree with this. Those who show strong concerns for their local environments are often criticised for acting from the Not-in-My-Backyard (NIMBY) syndrome. Focusing on local environments may obscure our vision of the global picture. We may be lulled, by relatively good local environmental standards, into forgetting the state of global environmental damage. In any event, attempts to preserve local environments may be undermined by deterioration in the supporting global environment: paradoxically, 'think locally, act globally'.

Although environmental law does not specifically identify 'local environments' as entities of concern, there are, nevertheless, aspects of law which reflect the view that local environments are conceptually important. One example of this is to be found in weight given to local concerns in the rules governing the right to challenge a decision taken by a public body – what lawyers call the right of 'standing' or *locus standi*. In *R* v. *Inspectorate of Pollution ex parte Greenpeace Limited* (No. 2) [1994] Greenpeace UK sought judicial review of a decision to grant a discharge consent to British Nuclear Fuels (BNFL) to bring their THORP reprocessing plant onto line. The judge, Otton J, in deciding that the applicants did have 'standing' to challenge that decision, commented:

> The fact that there are 400,000 [Greenpeace] supporters in the United Kingdom carries less weight than the fact that 2,500 of them come from the Cumbria region. I would be ignoring the blindingly obvious if I were to disregard the fact that those persons are inevitably concerned about (and have a genuine perception that there is) a danger to their health and safety from any additional discharge of radioactive waste even from testing.

A second aspect of law giving weight to the idea of *local* environments is the delegation of environmental responsibilities to local authorities. Planning law – also termed 'development control' – is a major local authority responsibility in the UK, and exerts a very significant effect in preventing undesirable environmental damage through development. Other areas of environmental responsibility for local authorities include the regulation of statutory nuisances under Sections 79 and 80 of the

Environmental Protection Act 1990 and the regulation of air pollution from smaller industrial polluters under Part I of that Act.

If it is desirable that the world should be conceptualised as multiple local environments then it is legitimate to ask whether we, as geographically located individuals, can be legitimately concerned with other societies' environments. This question has legal resonance in cases in which one nation-state seeks to influence, protectively, aspects of the environment outside its own territorial jurisdiction. However well intentioned, concern with the environments of other countries may be viewed as meddling interference or, at worst, environmental imperialism. In the case *United States – Restrictions on Imports of Tuna* (1991) a 'panel' (equivalent to a court) of the General Agreement on Tariffs and Trade was asked to decide whether an embargo on tuna imports by the United States was consistent with the GATT free trade rules. The US had embargoed tuna imports from Mexico because Mexico allowed purse-seine fishing (which results in dolphin deaths) in the Eastern Tropical Pacific Ocean. Generally the GATT convention does not allow parties to that convention to impose trade barriers between each other. However, exceptionally, Article XX(b) and (g) of the GATT allows trade barriers if necessary to protect human, animal or plant life or health or if aimed at the conservation of exhaustible natural resources respectively. The panel reasoned that the US embargo was not lawful since, in its view, Article XX(b) and (g) only permits a nation-state to protect life and health or exhaustible resources if these are located *within its own territorial jurisdiction*. A later GATT panel, *Dispute Panel Report on US Restrictions on Imports of Tuna* (1994), again concerning US tuna embargoes, reached the opposite conclusion on this point, illustrating confusion over the status of non-national aspects of the environment.

The issue of local environments versus one global environment has also arisen in US Environmental Impact Assessment (EIA) law. Federal cases give conflicting answers to the question of whether EIA should apply to US projects taking place outside of US territory. The court in *Environmental Defence Fund* v. *Masey* (1993) held that the US National Science Foundation was required to carry out an EIA before incinerating waste in Antarctica. *Greenpeace USA* v. *Stone*, however, ruled that an independent EIA is not required in advance of shipments of hazardous chemicals to US army bases in the Pacific Ocean. Kibel (1999) argues that the resolution of this inconsistency lies in recognising that NEPA exists to give US citizens a voice in determining projects that have environmental impacts. This principle applies wherever those projects may be located: to deny that voice in relation to overseas projects is to set double standards and to create a dangerous democratic vacuum. This is akin to supporting the notion of one global environment, in relation to which US projects must be assessed for environmental impact.

In EU law the question of the legitimacy of action to protect extraterritorial environments arises between Member States, and also between the EU as a whole and the rest of the globe. In *The Red Grouse Case* [1990] the European Court of Justice ruled that it was not lawful for the Netherlands to prohibit the sale of red grouse imported from the UK in Dutch shops: the Netherlands cannot impose its own environmental standards on another EU Member State. In relation to environment outside of the EU the Maastricht Treaty modified the EC Treaty to include, as an objective of Community policy, 'promoting measures at international level to deal with regional or world-wide environmental problems' (Art. 175). This makes it clear that the EU has competence to take steps to protect extra-European environments such as the ozone layer or the high seas.

The above issue – whether one can legitimately seek to *protect* the environment 'belonging' to others – is reversed in the question whether local environments can legitimately be *degraded* in pursuit of some global environmental benefit. For instance, s. 54A of the UK's Town and Country Planning Act 1990 creates a presumption that planning applications should be determined in line with local development plans, unless material considerations indicate the contrary. However, central government policy (PPG1, para. 30) advises that a local project may depart from the local development plan if 'the particular contribution of that proposal to some local *or national need or objective* is so significant that it outweighs what the plan has to say about it' (emphasis added). A concrete example of this might be a decision to allow construction of a wind farm which would create local environmental disturbance, but assist the national objective of reducing global warming (see *North Devon D.C. v. West Coast Wind Farm* (1996)).

Species and ecosystems

Should the concept of environment include abstract entities such as 'ecosystem', 'habitat' and 'species'? In favour of recognition and inclusion of these group or holistic entities is the argument that the whole is greater than the sum of the parts. When the final individual member of a species dies, and the species thereby becomes extinct, it is arguable that we have lost more than just the last few plants or animals. The loss also includes the *type* of organism and its genetic and physical characteristics.

Inclusion of ecosystems and species in the concept of environment is not entirely unproblematic. To begin with the very idea of 'ecosystems' is contested. Tansley (1935) developed the term in response to perceived weaknesses of the preceding holistic term – the 'biotic community' – which itself had originated in the work of the American plant ecologists F.E. Clements and John Phillips. Specifically, Clements and Phillips had imbued the 'biotic community' with characteristics of an organism, including, importantly, the tendency to develop through stages ('succession') to reach a stable 'climax community'. Tansley (1935) objected to this trend. He denied that nature's complexes are in fact organisms and he doubted the extent to which such communities proceed through succession to reach stable climaxes. Instead, Tansley offered a view of nature as organised into systems: *ecosystems*, a term which carries no purposive or teleological connotations and which avoids comparison with individual organisms.

The concept of the ecosystem has not gone unchallenged. Until the 1980s ecologists, influenced especially by Eugene and Howard Odum (1959), regarded ecosystems as stable equilibrium formations, more or less robust in the face of external (e.g. human) disturbance. However, in the late twentieth century this idea receded and the view that nature is unpredictable or chaotic, and in a state of permanent disequilibrium emerged (Bosselman and Tarlock 1994). In addition to these internal conflicts, not everyone agrees that ecosystems are 'real' entities. First, there is the old ontological objection that ecosystems are no more than the sum of their parts (Schindler 1987). A related objection is that an ecosystem is too large a unit to explain or to help understand nature: it is possible that 'the real action' takes place in smaller ill-defined units termed 'patches' (Pickett and White 1985). We can add to this the observation that convergence in succession may be more a statistical rather than a biological property (Brennan 1995).

Species, the other main 'whole' which environmentalists consider deserves legal protection, are usually defined as 'groups of interbreeding natural populations that are reproductively isolated from other such groups' (Mayr 1969, 311). Care is needed in relating morphological distinctness to reproductive isolation. Sometimes reproductive barriers do not lead to morphological variation (e.g. different species of fruit fly, outwardly indistinguishable). Conversely, unusual forms – the objects of great pleasure and interest for naturalists – are often 'mere' *varieties* of the same species. For example, the northern spotted owl interbreeds with and is therefore a mere variant of the more prevalent Californian spotted owl (Grierson 1992).

There are difficulties with the biological species concept. There is the usual ontological objection – i.e. that nature consists only of individuals. Then there is the problem of *semi-species*: sometimes individuals in group A can interbreed with those in group B, but not with those in group C, although individuals in group B can interbreed with those in group C; are these individuals all one species? Individuals in two groups may be interfertile but also intersterile (such as the horse and donkey, producing the infertile mule). Third, it should be noted that 'species' might be a minority classification: the vast majority of organisms on earth engage in asexual reproduction (lizards, insects, crustaceans and bacteria) and, therefore, cannot be fitted into any particular closed gene pool.

Are 'ecosystems' and 'species' the full set of holistic entities that we should incorporate into 'environment'? Perhaps 'rivers', 'valleys' and 'forests' are abstract units deserving legal recognition. In a famous essay, 'Should Trees have Standing?', Stone (1972) argued for the right of valleys and lakes to be represented in court – although his proposition was rejected by the US Supreme Court (*Sierra Club* v. *Morton* (1972)).

Schrader-Frechette and McCoy (1994) emphasise the extent to which our predetermined *values* about what is good and worthy of protection in nature shape and inform the wholes that we find and describe in the natural world. De-Shalit (1997) maintains that our conceptions of the environment, including notions such as 'ecosystem', are often *politically* loaded: environmental theorists attach meanings to terms such as 'environment' which hide *normative political* claims about human society. A rich vein of environmental sociology contends that, to a large extent, 'nature' and 'environment' are social constructions (see Merchant 1987; Hannigan 1995, ch. 6).

And yet generally, whatever objections may be raised at a theoretical level, we do think of species, ecosystems and habitats as constituents of the environment. And when we use these words we mean something more than just the arbitrary collection of individual organisms and non-living components that are found there. We imply (and presume) that these things have some sort of collective reality. It is not surprising, then, to find legal concepts of the environment which refer to ecosystems and species. Section 2(1) of New Zealand's Resource Management Act is an excellent example:

> The 'environment' includes
> (a) Ecosystems and their constituent parts, including people and communities; and
> (b) All natural and physical resources; and
> (c) Amenity values; and
> (d) The social, economic, aesthetic, and cultural conditions which affect the matters stated in paragraphs (a) to (c) of this definition or which are affected by those matters.

Similarly, according to the Australian Environment Protection and Biodiversity Conservation Act 1999, s. 528:

> environment includes:
> (a) ecosystems and their constituent parts, including people and communities; and
> (b) natural and physical resources; and
> (c) the qualities and characteristics of locations, places and areas; and
> (d) the social, economic and cultural aspects of a thing mentioned in paragraph (a), (b) or (c)

Key points

- The legal concept of the environment certainly includes the three media: land, air and water. The extent to which each of these should be included is, however, a matter for debate.
- Definitions of the environment implicitly limit or extend the protective reach of environmental law.
- A critical issue is the extent to which the environment should be thought of as global or local.

Further reading

Grierson, K.W. (1992) 'The Concept of Species and the Endangered Species Act', *Virginia Environmental Law Review*, vol. 11, pp. 463–98. This contains thought-provoking analysis of what we mean by a species and what it is that US conservation laws are trying to protect.

Grieder, T. and Garkovich, L. (1994) 'Landscapes, the Social Construction of Nature and the Environment', *Rural Sociology*, vol. 59, p.1 and Simmons, I.G.

(1993) *Interpreting Nature: Cultural Constructions of the Environment*, London: Routledge. These two sources provide thought-provoking consideration of our cherished conceptions of the environment, within and without law.

Tribe, L.H. (1974) 'Ways Not to Think about Plastic Trees: New Foundations for Environmental Law', *Yale Law Journal*, vol. 83, 1315–48. A truly seminal early paper – raises many important issues that law, ethics and politics have still to address adequately.

Roberts, L.D. (1997) 'Environmental Standards For Prospective Non-Terrestrial Development' (http://www.permanent.com/archimedes/EnvironmentArticle.html). Professor Roberts proposes that, for celestial bodies, the law should (1) institute a cogent, private property system, (2) promote an integrated liability standard for private as well as public entities, (3) make use of existing civil judicial mechanisms, and (4) establish a tailored method for the creation of limited 'traditional' environmental regulation.

Discussion questions

1 Do you think that the legal concept of the environment should include:

  (a) plastic trees 'planted' along roadsides (see Tribe: 1974)?

  (b) humans?

  (c) industrial landscapes?

2 Which environmental entities are most important?

3 Is it best to construct laws around the notion of one global environment or many localised environments?

4 Should our legal concept of the environment stretch to include the surface and atmospheres of celestial bodies such as the moon and the planet Mars?

3 Sources of environmental law

- International law
- Regional law
- National law

This chapter introduces the three main sources of law – international, regional and domestic law – and discusses how each varies in effectiveness as an instrument of environmental protection. Each of the three sources has, as we will see, characteristics, often related to the methods of law formation or enforcement, which limit or advance its efficacy.

At the international level 'law' is very different in nature from the bodies of rules and principles recognised at national level. International law consists, to a significant extent, of the voluntarily negotiated agreements of sovereign states. As such it is predominantly reflective of state practices and aspirations. Whether these can provide sufficient momentum to halt ecological degradation is, as we shall see, unclear.

Regional law is of growing importance. For the purposes of this chapter we examine the environmental law of the European Union. This, due to its method of formation, and the supranational nature of the EU itself, is capable of binding European states to achieving environmental standards. EU law may, thus, have a 'forcing' or prescriptive capacity for enhancing environmental protection. In this respect EU law can be seen to be superior to international law. However, EU environmental law suffers from difficulties of implementation and enforcement which are at least as serious as those affecting international and national regimes.

National law is still the most prevalent source of environmental regulation, although often shaped by and giving effect to international and regional measures.

International law as a source of environmental law

Many environmental problems (e.g. climate change, acid rain, water pollution) are global or, at least, transnational in their range of occurrence. This fact alone makes international law a source of prime importance for the protection of nature.

The forms of international environmental law

The question, 'what is international law?' must first be addressed. It is important to observe that most international law is *inter-national*, i.e. concerns the relations of sovereign states one with another, rather than individuals with states or other individuals. This can sometimes work against those who seek to assert environmental rights against private bodies or corporations. For instance, in the US case *Beanal* v. *Freeport-Moran, Inc.* (1997) Beanal, the leader of an indigenous Indonesian group, claimed that Freeport, a mining company, had subjected his people to acts of grave environmental harm. In its judgment the court observed that this legal wrong requires state involvement for its commission. Beanal, unfortunately, had failed to allege the facts necessary to establish any element of *state* action in Freeport's environmental practices. Consequently, the court was unable to find Freeport Corporation liable for any breach of international environmental law.

One aspect of international law that is increasingly important to individuals in a modern global economy is the law of 'choice of forum'. Environmental damage in developing countries may be caused by firms whose parent companies or headquarters are located in developed countries thousands of miles away. If individuals who suffer from the environmental damage caused by such firms can sue in the developed state, they often stand a much better chance of obtaining justice.

The body within the United Nations organisation charged with responsibility for determining disputes in matters of international law is the International Court of Justice (ICJ). Article 38 of the statute establishing the ICJ lists the sources of international law which the ICJ may apply:

> (1) The Court, whose function is to decide in accordance with international law such disputes as are submitted to it, shall apply:
> (a) International conventions, whether general or particular, establishing rules expressly recognised by the contesting states;

Box 3.1

The Cape Asbestos litigation – *Lubbe and Others* v. *Cape Plc*, House of Lords (2000)

In South Africa there has been a great deal of asbestos-related disease caused by exposure to asbestos fibres in mining and manufacturing-related activities or as a result of proximity to asbestos mine tailings dumps. More than half of former asbestos mineworkers in the Northern Cape area suffer from lung diseases caused by asbestos, along with as many as one in 12 of the general population in certain mining areas. The principal disease is asbestosis – a debilitating and often fatal condition. Some of the mines are owned and operated by a British-based company: Cape Asbestos plc. Cape, which withdrew from South Africa, paid compensation to its British workers who had suffered lung disease but refused to pay compensation to around three thousand South African workers who claimed to have suffered asbestosis as a result of their employment.

After failing to receive compensation from the company a number of Cape's South African mineworkers filed a suit in the United Kingdom seeking damages. The workers' main argument was that the Cape's lack of presence (and assets) in South Africa, coupled with the non-availability of legal aid in that country, meant that it was very unlikely that a claim brought before South African courts would result in justice. Cape, on the other hand, maintained that the rule of 'forum non conveniens' should be applied, which means that the case should be heard in a different country (in this case, South Africa). The case reached the House of Lords where Bingham LJ referred to the established precedent *Sim* v. *Robinow* (1892) 19 R. 665 (1 Div) which established that

> the plea [that the case should be brought in another country] can never be sustained unless the court is satisfied that there is some other tribunal, having competent jurisdiction, in which the case may be tried more suitably for the interests of all the parties and for the ends of justice.

In relation to the asbestos litigation in hand he concluded:

> If these proceedings were stayed in favour of the more appropriate forum in South Africa, the probability is that the plaintiffs would have no means of obtaining the professional representation and the expert evidence which would be essential if these claims were to be justly decided . . . This would amount to a denial of justice.

Lord Bingham pointed to the fact that Cape had not objected to the hearing of a separate claim four Italians had brought against the company for asbestos-related illness.

Leigh, Day and Company, the law firm fighting the case of the three thousand Cape plc workers and their families, has subsequently launched cases against several other English asbestos mining companies with former overseas operations. The eventual damages are expected to be over R300 million.The Lubbe ruling creates a precedent for other claimants in other environmentally caused illness cases who are unable to fight their cause in their own countries for similar reasons. In future the transnationalness of companies will be used by those who suffer from environmental illnesses as an important strategy in the battle for compensation.

(b) international custom, as evidence of a general principle accepted as law;
(c) the general principles of law recognised by civilised nations;
(d) subject to the provisions of Article 59, judicial decisions and the teachings of the most highly qualified publicists of the various nations, as a subsidiary means for the determination of rules of law.

Although not intended to be a comprehensive list, Article 38 nevertheless reflects the most important divisions of international law. In the following exploration an additional class is added – so-called 'soft law': law which is morally and politically persuasive only.

Soft law

Lawyers often distinguish 'soft law' from the main 'hard law' sources of international law. Soft law consists of statements – resolutions, recommendations, declarations, 'final acts' of conferences and similar documents – that are strongly persuasive but not, in a strict sense, legally binding (Palmer 1992).

Given their cultural, economic, and social diversity, states are more likely to agree to soft law than to legally binding instruments such as conventions. As such, soft law can provide an intermediary step between mere policy statements and fully binding legal commitments and thus facilitates the *incremental* movement of states towards stronger environmental commitments. Many important instruments of 'hard' international environmental law have their origins in soft law. For instance, the United Nations Convention on the Law of the Sea 1982 and the 1979 Moon Treaty both contain statements to the effect that the surface of the seabed and the moon are part of the 'common heritage' of mankind; statements that can be traced to a non-binding declaration, United Nations General Assembly Resolution 2749, which stated that 'The sea-bed and ocean floor, and the subsoil thereof, beyond the limits of national jurisdiction . . . are the common heritage of mankind.'

Soft law's contribution to environmental protection is often one of repetition of commitments, cross-referencing from one body or instrument to another, until a common international understanding develops (Dupuy 1991). However, even within these multiple references, certain key statements or understandings predominate. Amongst the most significant environmental soft law instruments are the 1972 Stockholm

Declaration on the Human Environment (Sohn 1973) and the 1992 Rio Declaration on Environment and Development (Grubb *et al.* 1993).

The principles of the Stockholm Declaration did not create any specific binding legal obligations (*Amlon Metals, Inc.* v. *FMC Corp.* (1991)). They nevertheless served to fire the global environmental movement and marked a turning point in world consciousness of environmental problems.

The 1992 Rio Declaration, which incorporates and updates most of the principles of the Stockholm Declaration, is one of the five concrete outcomes of the 1992 United Nations Convention on the Environment and Development (UNCED). Key themes of the Rio Declaration are the principles of 'common but differentiated responsibilities' (Principle 7) and state sovereignty (Principle 2) (Porras 1993).

'Common but differentiated responsibility' means that developed states have special responsibilities; 'in view of the pressures their societies place on the global environment and of the technologies and financial resources they command'. The principle of state sovereignty refers to the 'sovereign right of states to exploit their own resources pursuant to their own environmental and developmental policies'.

The 27 principles of the Declaration, summarised in Table 3.1, can either be regarded as a delicate balance between developmental and environmental interests (Nanda 1995a) or, because of their emphasis on development and sovereignty, as a retrograde step for environmental protection (Pallemaerts 1993).

Conventions

Conventions are the most important form of international environmental law. The term 'convention' includes treaties, statutes, protocols and any other form of express written agreement concluded between states. Occasionally, even a unilateral statement issued by a state can be regarded as equivalent to a convention if it is worded so as to create a legal obligation. Such cases will, however, be narrowly construed. In the nuclear tests cases (*Australia* v. *France, New Zealand* v. *France* (1974)) the ICJ regarded an assurance by France in 1974 to cease atmospheric nuclear testing as legally binding on that country. In 1995 New Zealand objected to the ICJ regarding France undertaking nuclear testing on the Mururoa and Fangataufa atolls without first conducting prior

Table 3.1 ***1992 Rio Declaration on Environment and Development***

Article	*Summary*
1	A human right to a 'healthy and productive life in harmony with nature'
2	States' right to exploit own resources
3	Development for developmental/environmental needs of present/future generations
4	Environmental protection to an integral part of the development process
5	Eradication of poverty 'an indispensable requirement for sustainable development'
6	Special situation and needs of developing countries to be given special priority
7	States have 'common but differentiated responsibilities' for environmental protection
8	Reduce unsustainable production/consumption; appropriate demographic policies
9	Exchange of scientific and technological knowledge and technologies
10	Environment issues require participation of all concerned citizens
11	States shall enact effective environmental legislation . . .
12	States to cooperate to promote a supportive and open international economic system
13	Provide liability/compensation for victims of pollution and 'other environmental damage'
14	States to discourage/prevent the relocation/transfer to other States of damaging activities
15	Precautionary approach to be widely applied by States according to their capabilities
16	The polluter pays principle to be applied through economic instruments
17	Environmental assessment for activities with significant adverse environmental impact
18	Notification of other states in the event of an environmental emergency
19	Provision of environmental information to States *re* transboundary pollution
20	'Women have a vital role in environmental management and development . . .'
21	Ideals and courage of youth to be mobilised to achieve sustainable development
22	Vital role of indigenous people in environmental management
23	Protection of environments of oppressed, dominated or occupied peoples
24	'Warfare is inherently destructive of sustainable development'
25	'Peace, development and environmental protection are interdependent and indivisible'
26	States to resolve environmental disputes peacefully and by appropriate means
27	States and people to cooperate in good faith to fulfil these principles

environmental impact assessment. A majority of the ICJ judges rejected New Zealand's claim that France would be breaking its 1974 undertaking, on the narrow ground that the tests concerned were not strictly atmospheric, hence not covered by the earlier French promise.

The great strength of conventions when it comes to environmental protection is the relative precision and clarity of the obligations that they impose on states. This precision is evident in conventions which provide 'built-in' lists of, for example, regulated substances or protected species. For example, the Convention on International Trade in Endangered Species (CITES) utilises three lists: Annexe I (species threatened with extinction); Annexe II (species not necessarily threatened with extinction but may become so) and Annexe III (species which any party identifies as being subject to regulation within its jurisdiction for the purposes of preventing or restricting exploitation, and as needing the cooperation of other parties in the control of trade). The degree of control over trade in species in each Annexe is correspondingly tight. Annexes can easily be modified to deal with increases in scientific knowledge or to reflect changes in social values. The CITES annexes have been frequently modified, and inclusion of new species onto Annexe I has acted as a driving force for other conservation measures for those species.

Other conventions achieve precision in a step-wise manner. A main convention is negotiated which is deliberately broad or 'framework' in nature. The detail of the regime is postponed for elaboration in subsequent and subsidiary conventions. These 'daughter' conventions are termed 'protocols'. The system is exemplified by the 1985 Vienna Convention on the Protection of the Ozone Layer. This contained the quite ambiguous obligation to take 'appropriate measures' to protect human health and the environment against 'adverse effects resulting or likely to result from human activities which modify or are likely to modify the ozone layer' (Art. 2). The subsequent 1987 Montreal Protocol to that convention crystallised a set of very precise obligations, including the duty to reduce consumption of five ozone-depleting substances (ODSs) to 50 per cent of 1986 levels by 1998. This protocol has been revised, bringing forward the phase-out timetable and greatly increasing the range of ODSs to which the obligations apply.

Conventions do have weaknesses as instruments of environmental protection. The most serious is that they are deemed to be binding only on those states that have indicated their agreement by signature and ratification. This limitation derives from Grotius's philosophy of the *consensual* nature of international law. This, in turn, derives from the 'rule' that states cannot, and should not, be regarded as subject to limitations on their sovereignty without first explicitly or implicitly indicating their consent to be so bound. Of course, not all commentators agree that the sovereignty of states should be rigidly adhered to when the integrity of the global environment, even survival of life itself, is at stake (McNair 1934; Palmer 1992). The World Commission on Environment and Development (WCED), for one, has doubted the

> feasibility of maintaining an international system that cannot prevent one or several of the states from damaging the ecological basis for development and even the prospects for survival of any other or even all other states.
>
> (WCED 1987, 313)

Hey (1992), in similar vein, comments: 'Perhaps the time has come for a more balanced approach to state sovereignty, especially where environmental issues are concerned.'

The above propositions might be realised if certain rules of international environmental law were considered to be law *ergo omnes* (i.e. good against the whole world). Currently, only those rules necessary for the survival and operation of the global system of sovereign states itself – e.g. treaties containing rules concerning territorial boundaries – have this status. But sovereign states and their inhabitants are, in turn, dependent on the continuation of the global biotic life support system. It would not, therefore, be unrealistic to consider that, in the future, rules concerning the protection of critical natural capital (e.g. the ozone layer) may be considered binding on all states whether or not they consent.

A partial solution to the problem of states' avoidance of conventions is the notion of 'treaty incorporation'. This occurs when agreement to one convention automatically implies agreement to another or others. Article 211 of the 1982 United Nations Convention on the Law of the Sea (UNCLOS), for instance, incorporates the precise standards for operational oil pollution set by the earlier 1973/78 MARPOL convention. All states that agree to be bound by UNCLOS are also bound by the MARPOL standards.

States may be bound against their consent to changes in conventions or protocols to which they have become parties. This may occur due to terms of the convention that provide for the use of a majority vote for decisions and the promulgation of detailed rules. A clear instance of such an approach is Article 2(9)(c) of the Montreal Protocol on Substances that Deplete the Ozone Layer of 1987. This provides, in relation to *adjustments* to that protocol, that when efforts to achieve consensus fail, binding decisions may be taken by a two-thirds majority vote of parties representing at least 50 per cent of the total consumption of ozone-depleting substances.

A second limitation of conventions is that, despite their relative clarity and precision, they often contain substantive obligations that are weak and insufficient to halt the environmental deterioration in question.

Sometimes the weakness derives from the use of ambiguous or vague language in the convention. Illustrative in this respect is *The Commonwealth* v. *The State of Tasmania* (1983). This case arose from plans, by Tasmania, to construct a dam on the Gordon River in an area which was a designated site on the World Heritage List, set up under the World Heritage Convention. Australia issued an order protecting the site, making the dam illegal. Tasmania challenged the legality of this order before the Australian High Court, arguing that Article 5 of the Convention did not create a *legally* binding obligation to definitely protect areas on the list. Article 5 spoke only of the need for party states to 'endeavour, in so far as possible, and as appropriate for each country [to take various protective steps]'. Fortunately, in this case a majority of the High Court thought that, despite the weak wording, Article 5 did create a legal obligation. However, the case illustrates the problems associated with the rather loose wording of many conventions.

Lack of substantive rigour in conventions may result from insufficient scientific understanding of the problem in question or, more commonly, from political resistance at the time of negotiation by states with vested interests. It is generally agreed that the weakness of the obligations in the 1992 United Nations Framework Convention on Climate Change is partly a result of reluctance on the part of the United States to commit to major internal changes.

Finally, conventions often fail to provide adequate environmental protection because of the availability of objection and reservation procedures. These can take two forms: joining reservations or continuing reservations. In the first case, states may avoid convention obligations by entering a qualification or reservation at the time of their ratification of the convention in question (e.g. CITES, Article 23(2)). In the second case, states can escape from *new* rules or lists created under environmental conventions. Article VI of the International Convention for the Regulation of Whaling 1946 provides a 60-day period during which party states can register objections to any changes to the catch quotas set under the convention's Annexe. This has been used many times (e.g. by Norway, USSR, Iceland and Japan in 1983) to avoid reductions in catch quotas. Similarly, Articles 25(3) and 26(2) of CITES allow reservations to be entered to subsequent changes to the Appendices within 90 days of any such change. Some species, such as turtles and tortoises, have been badly affected by such actions (Lyster 1985, 263). Some commentators see these 'get-out clauses' as positive features: 'the ability of a regime to bend, but not break gives it strength and resilience over time and allows for deeper commitments' (Victor *et al.* 1998, 693). Others are more sceptical.

Customary law

The relations between states, in so far as they are not set out in written agreements, are governed by customary law: a body of law which reflects the 'settled practice' of states. Customary law requires two things: (a) evidence of consistent state practice, and (b) evidence that states recognise such practice as legally binding upon them. The latter element is called *opinio juris*. Customary law automatically forms part of the national law of most states (e.g., in the UK, *Trendtex Trading Corporation* v. *Central Bank of Nigeria* [1977]).

The great strength of customary law is that, unlike conventions, it is binding on all states. The elements of customary international environmental law include:

- a requirement of prior consultation in matters affecting transboundary resources (*Lac Lanoux Arbitration* (1957));
- a requirement to warn other states in the case of transboundary environmental hazard (*Corfu Channel Case* (1949));
- a requirement to give equivalent legal treatment to domestic and non-domestic pollution (Birnie and Boyle 1992, 111); and
- a requirement to engage in equitable utilisation of shared resources (Birnie and Boyle 1992, 219).

In theory, states can escape from customary law by persistently and originally 'opposing' an emerging practice (e.g. *The Anglo-Norwegian Fisheries Case* (1951)). Opposition could be relied upon by states wishing to resume commercial whaling or by those which prefer not to be bound by the duty to future generations and the Precautionary Principle (McIntyre and Mosedale 1997). So far, fortunately, states have only infrequently made use of this feature of international law.

The most significant disadvantage of customary law, from the environmental perspective, is its relative lack of certainty and clarity. For example, in the matter of transboundary harm – e.g. acid rain or radioactive fallout – customary international law obliges states to prevent their territory from being used to cause damage to the territory of other states. Yet the precise extent and content of this obligation is far from clear. It is generally interpreted in a rather weak sense as requiring no more than the exercise of 'due diligence' to prevent actual harm – possibly only foreseeable and serious harm (Birnie and Boyle 1992, 96).

Other sources of international environmental law

Other sources of international law, as mentioned by Article 38 of the ICJ statute, include general principles of law and, as subsidiary sources, judicial decisions and the writings of eminent publicists.

General principles are usually thought of as comprising principles common to all western legal systems. An example might be the principle that both sides of the dispute must be heard before the court takes a decision, and the principle that one should not use one's property so as to cause injury to a neighbour (nuisance law). It is possible, although by no means certain, that 'declarations of principles' such as the UN declaration of the World Charter for Nature may be considered to be part of international law. We shall see in Chapter 4 that norms such as the Polluter Pays Principle, the Precautionary Principle and the Principle of Sustainable Development are part of, or rapidly becoming part of, international law.

Creating international environmental law

The processes leading to the formation of the various elements of international law differ significantly. Conventions are, theoretically, the result of the exercise of the free will of sovereign nations. The impetus for a new convention may arise from some political forum. Alternatively, the impetus may arise from an existing convention which has deliberately left difficult or technical features unresolved. Article 211 UNCLOS, for instance, requires states to

> establish international rules and standards to prevent, reduce and control pollution of the marine environment from vessels and promote the adoption, in the same manner, wherever appropriate, of routing systems designed to minimize the threat of accidents which might cause pollution of the marine environment.

The process of convention formation has several stages. First, a draft convention is drawn up under the auspices of some international or regional body such as the United Nations Environment Programme (UNEP) or the Council of Europe (COE). This acts as an object of discussion and a focus for negotiations. Meetings then take place between the governments of interested states according to an agreed negotiating process, usually as part of a 'conference' or series of conferences. During the discussions amendments will be made to the original text. Parties finding themselves still in agreement with the negotiated instrument will then *sign* the convention. Signature is not, however, the end of the matter: *ratification* by a set number of states (less commonly, by states representing a given percentage or amount of the regulated activity) is required before the convention has legal effect. The multiple ratification criterion is designed to avoid prejudicing states which ratify early, but it can also have the effect of imparting considerable delay to the creation of international law. Sometimes delay turns to abandonment, as was the case with the Convention on the Regulation of Antarctic Mineral Resources. This convention was agreed to in 1988 but never came into force due to the gradual emergence of a conviction that its provisions were insufficient to protect the Antarctic region.

Although the process of ratification consists merely of deposition of an instrument of ratification with the relevant convention secretariat, ratification may, depending upon the nature of the state in question, entail the passage of new national law translating the requirements of the relevant convention. This final step is not necessary in 'monist' states, such as the United States, which consider international and domestic law to be one undivided body of rules. It is essential, however, in 'dualist' states such as the UK, Canada, Australia and India which consider the two to be separate domains (e.g. *Maclaine Watson* v. *Dept of Trade and Industry and related appeals* [1989]).

Many factors affect the willingness of states to sign and ratify environmental conventions. There may be obvious issues of self-interest, as in conventions dealing with resource conservation (e.g. fisheries measures). States may wish to satisfy their domestic 'green electorate' or to increase their international kudos. Other positive reasons for agreement include the prospect of receipt of funding or technological assistance. Article 20(2) of the 1992 Biological Diversity Convention, for instance, encourages ratification by developing states:

> The developed country Parties shall provide new and additional financial resources to enable developing country Parties to meet the agreed full incremental costs to them of implementing measures which fulfil the obligations of this Convention.

States may also be attracted by the possibility of shared benefits. Article 19 of the 1992 Biological Diversity Convention, for instance, requires states to take

> all practicable measures to promote and advance priority access on a fair and equitable basis by Contracting Parties, especially developing countries, to the results and benefits arising from biotechnologies based upon genetic resources provided by those Contracting Parties.

Political pressure may be exerted on unwilling parties by powerful states such as the United States to persuade them to join. Peru and Chile were notably 'persuaded' by the US to join the International Convention for the Regulation of Whaling and, thereby, fall within the regime of the IWC. Economic pressure can be created by insertion of economic barriers to trade between participating and non-participating states, as is the case with the Vienna Convention on Ozone Depletion which prohibits parties to the convention from trading in ozone-depleting substances with non-parties.

Customary international law is, as mentioned, formed from the settled practice of states coupled with a sense of legal obligation or right – *opinio juris*. The difficulty with this form of law is ascertaining when sufficient state practice and *opinio juris* have accumulated. As far as 'consistent state practice' is concerned, the practice must be fairly general or widespread. Possibly a rule imposing positive obligations on states requires to be more widespread than one merely requiring states to allow certain practices (e.g. Dixon 1990, 26). 'Practice' can include acts and omissions, statements, legislative acts, indeed almost any kind of state activity. The practice in question may begin with only a few states and then 'snowball'. For instance, in 1945 President Truman issued a statement claiming that the United States owned the resources of the seabed adjacent to its coast. Other states soon followed suit. This led to a general belief that claims over large reaches of coastal waters were legally permissible under international law. Thus arose an element of customary marine law. Incidentally, this particular rule completed the transition from customary law to convention when extensive coastal state 'ownership' of continental shelf resources was included in the 1958 UN Convention on the Continental Shelf. It is also now found in the UNCLOS.

Unfortunately, *opinio juris* cannot simply be inferred from consistent state practice itself (*North Sea Continental Shelf Cases* (1969)). Since states often do not say whether they are or are not acting out of a sense of legal obligation/right it is usually difficult to be sure whether the necessary *opinio juris* has formed. Often, *opinio juris* must be inferred from statements or related actions of states. In *Nicaragua* v. *United States* the ICJ felt able to use the states' attitudes towards UN General Assembly resolutions as the basis for *opinio juris* – in this case the *opinio juris* of a rule of customary law that states must generally refrain from the threat or use of force against other states.

Enforcing international environmental law

Enforcement of international environmental law is vital yet problematic. As Boyle (1991) comments:

> [t]he development of rules of international law concerning the protection of the environment is of little significance unless accompanied by effective means for ensuring enforcement and compliance.
>
> (Boyle 1991, 229)

Fortunately, states usually comply with international environmental law. There are several reasons for such behaviour:

- Nothing may be lost by complying: for example the Antarctic Treaty's prohibition on mining relates to an activity which no state would currently wish to engage in, for straightforward economic reasons (Mitchell 1996).
- Compliance may be, in part, due to ambiguous language used in the law in question: allowing compliance with minimal disruption to state activities.
- Compliance may reflect practices already occurring or already anticipated by the state concerned. During transition from a socialist to a capitalist economy Russian compliance with sulphur dioxide and nitrous oxide protocols was easily obtained through the natural shrinkage in industrial activity that accompanied that transition.
- Compliance may serve internal political goals; for example, appearing to take a tough stance on the environment or needing some justification for imposing higher costs or taxes. Compliance may satisfy broader notions of 'interest'.
- States may seek to raise their status in the international 'league of nations' and, conversely, may be keen to avoid a bad reputation. Non-compliance in one area casts in doubt the ability of a state to 'deliver' in another, thus undermining future negotiating strength.
- The treaty in question may be backed by powerful nations such as the United States – as is the case with the International Convention for the Regulation of Whaling.
- Non-compliance on one treaty may lead to fears of retaliatory measures (including non-compliance) by other states on that or other treaties.

A variety of mechanisms tend to improve states' implementation and enforcement:

- Requirements to conduct environmental impact assessments (EIAs); reporting and information provision, and terms imposing liability for environmental damage (Sands 1995).
- High levels of public participation and government accountability, typically found in liberal-democratic states (Slaughter 1995).
- Institutional machinery and secretariats created by conventions, gathering data and monitoring implementation (Boyle 1991).

Although states may comply, partial or delayed compliance is not infrequent. Rest (1998, 64) notes 'a huge deficiency in implementation [of environmental conventions] and a lack of mechanisms for enforcement'. Nanda (1995a) analyses the problem thus:

> The state-centred nature of the contemporary international arena lies at the root of the problem. Compliance is voluntary as a result of the horizontal nature of the international system. Thus there is little incentive to comply when non-compliance is perceived as better serving national interests.
>
> (Nanda 1995a, 6)

Non-compliance may also be due to *inability* through lack of capacity, inadvertence, or some combination of both (Mitchell 1996). In respect of the former many modern environmental conventions seek to assist developing states, which may be unable to afford the protective measures necessary, or which may not have access to the necessary technologies. Article 4(5) of the 1992 Climate Change Convention requires developed country parties to

> take all practicable steps to promote, facilitate and finance, as appropriate, the transfer of, or access to, environmentally sound technologies and know-how to other Parties, particularly developing country Parties, to enable them to implement the provisions of the Convention.

What happens when non-compliance occurs? Failure to comply with national or regional law usually results in appearance before a national or regional court, but this does not often occur in the context of international environmental law. There is, as yet, no international environmental court – although some have argued for its creation (Rest 1998). States cannot be forced to come before the ICJ: submission to its jurisdiction is strictly voluntary. Due to the unpredictable and politicised nature of past ICJ decisions, states are generally reluctant to agree to that course of action:

between 1945 and 1986 only 72 cases were submitted, resulting in 45 judgments and 17 advisory opinions (Kubasek and Silverman 2000, 327). States thus prefer to settle their differences though diplomacy, private litigation or *ad hoc* arbitration (e.g. *The Trail Smelter Arbitration* [1941]).

After an ICJ ruling, the 'losing' state may be reluctant to alter its practices or pay compensation. The 1949 ICJ ruling that Albania should pay damages for damage caused to UK warships by mines laid in Albanian waters (*UK* v. *Albania* (1949)) was not satisfied until the 1990s. UN members undertake to obey the ICJ's ruling if they are a party to the case, but the court has no powers to ensure compliance with its judgments. The UN Security Council may decide upon 'measures to be taken to give effect to the judgment' (UN Charter, Article 94). Indeed, Rielly (1996) contends that UNSC action is appropriate in the case of global environmental harms that are 'both substantial and probably irreversible'. In reality, though, states know that the UNSC is unlikely to impose any such measures. ICJ decisions thus lack 'bite'.

The ICJ has a special environmental chamber, but this has not yet been used. Views on the ICJ's approach to environmental disputes are mixed, as evidenced by the recent *Case concerning the Gabcikovo-Nagymaros Project* (1997). This case arose from a dispute between Hungary and Slovakia over the lawfulness of Hungary's termination of work on a joint project for two barrages across the Danube, on environmental grounds. McGillivray and Mansell (1998) point out that the ICJ did not apply basic principles of environmental law such as the Precautionary Principle, preferring instead to utilise conventional 'black letter' modes of analysis. Sands (1998), on the other hand, observes that the court not only accepted the existence of a principle of 'ecological necessity' which could, in principle, justify the performance of otherwise unlawful acts, but also accepted the need for current standards to be used in the determination of environmental risk. The court has also commented, in its *Advisory Opinion on the Legality of the Threat or Use of Nuclear Weapons* (1996), on the importance of taking environmental factors into account in the context of the rules of law of armed combat. These cases indicate an increase in the court's sensitivity to environmental issues.

Despite calls for its creation (World Affairs Council of Philadelphia 1975; Allot 1990), or conversely warnings against the same (Jasper 1996), there is, as yet, no world government to press sanctions against a state which ignores international law. In its absence, the main instruments of compliance with international environmental law are (1) multilateral dispute resolution through institutional machinery; (2) pressure of public opinion and the world community; (3) litigation in national courts; (4) economic sanctions.

Regarding the first of these, almost every environmental convention creates some form of governing institution, the role of which includes supervision of implementation of the convention by party states and arbitration of disputes (Boyle 1991). Regarding the second, the embarrassment of an offending state by environmental NGOs publicising non-compliance often produces changes in government policy. Environmental NGOs may resort to litigation before national courts to force compliance with international law. For example, in the Dutch *Herpetologische Studiegroep* case, an international environmental agreement was successfully used to challenge a land use plan (Bodansky and Brunnée 1998).

Economic sanctions, although not common, have been employed by some states, notably the United States. Economic pressure was, for example, employed to bring Japan within the IWC whaling moratorium adopted by the IWC in 1982. Under the 1971 Pelly Amendment to the US Fisherman's Protective Act the Secretary of Commerce may certify that the activities of a state '[diminish] the effectiveness of an international fishery conservation programme' to which the USA is a party. Following such certification the President has 60 days in which to determine whether a portion, or all, or none of the fish products from the offending country shall be banned from importation. The US has threatened to impose trade sanctions several times: each time the threat alone has been sufficient to bring about an alteration in the 'offending' state's practices (Jenkins 1993)

Economic sanctions tend to be used only occasionally due to the threat of economic retaliation, the lack of impact in many cases, and their dubious legal validity in the light of the General Agreement on Tariffs and Trade (GATT).

Regional law

There are good reasons why much, indeed most, environmental law exists not at international level but at the level of regional or domestic governance. Many environmental problems, such as acid rain damage, are principally regional in geographical scope. Some matters of environmental concern are located entirely in the territory of one state and pose little risk to neighbours (e.g. contaminated land). Second, as we saw in the previous chapter, international conventions often require transposition into some form of domestic or regional arrangement in order to take effect in the internal legal systems concerned. Third, given the differences in environmental conditions that exist between regions it is appropriate that differing environmental laws and standards should apply in different regional areas.

Each major continent or world area has some form of regional law. Here we shall concentrate on the European Union's legal system, but also consider, briefly, American, African and Asian regional measures.

EU environmental law

Europe is unique in having created a supranational body, the European Union (EU), to which extensive legislative powers have been ceded by its Member States. The EU began in 1952 with the agreement, by France, Germany, Belgium, Italy, Luxembourg and the Netherlands, of the European Coal and Steel Community (ECSC). Further emphasis on economic integration resulted in the conclusion, in 1957, of treaties establishing the European Atomic Energy Community (Euratom) and the European Economic Community (EEC). Additions to the EEC include: Denmark, Ireland, the UK (1972); Greece, Spain and Portugal (1986); the former German Democratic Republic (1990 – through unification); and Austria, Finland and Sweden (1995). The EU now comprises around 368 million people and 15 Member States. Ten central European countries, plus Cyprus, Malta and Turkey, are currently seeking EU membership. They first need to fulfil political and economic requirements agreed in 1993 (the 'Copenhagen Criteria'). So far Malta and Cyprus are the only states to fulfil all economic and political criteria (CEC 1999b).

There has also been considerable structural development of the Community. The 1987 Single European Act (SEA) changed the EEC's name to the European Community (EC), thus emphasising a wider political purpose. The 'Community' underwent metamorphosis into a 'Union' under the 1992 Treaty on European Union (TEU), or 'Maastricht Treaty' as it is otherwise known. The European Union comprises the EC along with a Common Foreign and Security Policy (CFSP) and cooperation on matters of Justice and Home Affairs (JHA). The CFSP and JHA are predominantly matters of intergovernmental agreement and cannot, therefore, make law as the EC does.

European environmental law is a product of the EU's internal institutions:

- The Council of the European Union (http://ue.eu.int/en/summ.htm)
- The European Parliament (http://www.europarl.eu.int/home/default_en.htm)
- The European Commission (http://europa.eu.int/comm/index_en.htm)
- The European Court of Justice (http://europa.eu.int/cj/en/index.htm)
- The European Environment Agency (http://www.eea.eu.int/)

The Council is the EU's legislature. It consists of the government minister responsible for the topic under discussion from each Member State (i.e. every environment minister in relation to environmental proposals). The Presidency of the Council rotates every six months, enabling different states to set the legislative agenda. Council Decisions may be made in three ways: by simple majority; by qualified majority; or by unanimity. The Members States have differential voting strength, reflecting population levels, ranging from ten votes each for the UK, Germany, Italy and France, down to two votes for Luxembourg. The EC Treaty requires that most decisions relating to proposals for environmental matters are taken by qualified majority vote: currently 62 votes out of 87.

The European Parliament is a consultative and advisory body consisting of 626 MEPs directly elected for five years. Its main task is to scrutinise legislation, especially through the influential Environment, Public Health and Consumer Committee. In some cases it acts in concert with the Council in passing environmental legislation. It also controls the EU budget. Informally the Parliament sees itself as the champion of citizen and environmental rights. Its involvement in the passing of EU environmental legislation is, therefore, a good thing.

The European Commission, which comprises twenty commissioners nominated by the Members States for a five-year period, supported by a body of civil servants, has three functions: initiator of legislation; guardian of the treaties; and community executive. Internally the Commission is organised into sub-units – 'Directorate Generals' ('DGs') – including environment, agriculture, fisheries and energy and transport DGs.

The role of the European Court of Justice (ECJ), under Article 220 EC Treaty, is to ensure that EU law is applied. Environmental cases are normally either actions brought by the Commission alleging failure by a Member State to properly implement EU environmental law (Article 226 EC Treaty) or references made by national courts seeking clarification of the proper interpretation of EU law (Article 234 EC Treaty). In every case the dispute is first subject to the Opinion of an Advocate-General, drawn from a panel of six. These Opinions are highly influential and, more often than not, form the basis of the Court's decisions. ECJ Decisions and Advocate-General's Opinions are published in hard copy and Internet forms (http://curia.eu.int/en/). These are supplemented by independent compendiums of ECJ environmental decisions (Krämer 1993, and http://www.eel.nl/cases.htm) and commentary in journals such as the *Review of European Community and International Environmental Law and European Environmental Law Review.*

Although not technically an EU institution, a further body with importance for environmental protection within the EU is the European Environment Agency (EEA) established by EEC Regulation 1210/90. The EEA is not an environmental enforcer in a sense comparable to that of the UK's Environment Agency. Neither is it an environmental standard-setter like the US Environmental Protection Agency (EPA). Instead its main tasks consist of the gathering, analysis and dissemination of environmental data (http://service.eea.eu.int/seris/files/weur.shtml). It is likely that in the future its roles will be expanded, possibly including auditing of Member States' national environmental regulators.

The creation of EU environmental law

The EEC, as we have seen, was originally a predominantly economic organisation. However, its competence to create environmental law and policy has expanded considerably over the last thirty years. In 1987 the SEA provided the EC with explicit competence on environmental matters by inclusion of new articles 130 r–t into the EC Treaty. Article 130r, which was renumbered as Article 174 by the 1997 Amsterdam Treaty, sets out the basis for EU environmental policy:

> Community policy on the environment shall aim at a high level of protection taking into account the diversity of situations in the various regions of the Community. It shall be based on the precautionary principle and on the principles that preventive action should be taken, that environmental damage should as a priority be rectified at source and that the polluter should pay.
>
> In this context, harmonisation measures answering environmental protection requirements shall include, where appropriate, a safeguard clause allowing Member States to take provisional measures, for non-

> economic environmental reasons, subject to a Community inspection procedure.

Concrete EU environmental policies are contained in a number of 'Action Programmes', white papers and other discussion documents. The current (5th) action programme, *Towards Sustainability*, emphasises the need for a holistic approach to environmental problems (i.e. integration of the environmental dimension in all major policy areas), and utilisation of a broad spectrum of instruments such as laws, economic instruments, green technologies and education, for their resolution.

Specific proposals for items of EU environmental legislation are generated by an initial Commission proposal in the form of a white paper, or 'COM Doc', and also published in the EU's official journal. Simplifying the matter somewhat, after consultation with the European Parliament and the Economic and Social Committee, the final decision on whether to adopt a new EU environmental law is taken by the Council, although in certain cases the Parliament is given a power of 'co-decision'.

The legislative outputs of the Council – Directives, Regulations, Decisions, Recommendations and Opinions – have different legal effects. Article 249 EC Treaty states:

> A regulation shall have general application. It shall be binding in its entirety and directly applicable in all Member States
>
> A directive shall be binding, as to the result to be achieved upon each member state to which it is directed, but shall leave to the national authorities the choice of form and methods.
>
> A decision shall be binding in its entirety upon those to whom it is addressed.
>
> Recommendations and opinions shall have no binding force.

Directives, which form the mainstay of EU environmental law, are instructions to Member States and are not intended, therefore, to create rights and obligations for individuals. They require 'transposition' by enactment of national law in the Member States. This has the advantage of allowing the smoothest possible integration of new environmental laws into the diverse European legal systems. Regulations, which are less commonly used, are 'directly applicable' – i.e. they require no transposition into national law and do create rights and obligations for individuals. Regulations have the advantage of ensuring a totally unified and detailed regime across the Member States. Over three hundred items of environmental legislation have now been created (http://www.asser.nl/EEL/index2).

Krämer (1996) points out that EU environmental law has a number of distinctive characteristics, which are generally negative in character:

- Community institutions do not have a general competence;
- there is no Community-wide public opinion influencing the law-making process;
- regulations are adopted behind closed doors without public participation;
- the rules generally address Member States not individuals;
- implementation is normally done at Member State level;
- Community rules are largely ignored.

Enforcing EU environmental law

There are three groups or bodies which have some role to play in the enforcement of EU environmental law: the EU Commission, individuals, and EU Member States.

The key enforcement actor is the Commission which, in its role as 'guardian of the treaties', is charged with ensuring proper implementation of EU law. The record of Member States in implementation is not excellent: in 1995 Member States had implemented only 91 per cent of environmental Directives and the Commission registered 265 suspected breaches of Community environmental law (over 20 per cent of all the infringements registered by the Commission in that year). Over six hundred environmental complaints and infringement cases were outstanding against Member States in October 1996, with eighty-five of the latter awaiting determination by the ECJ (CEC 1997). Recent reports show similar trends (CEC 1999a).

If convinced that a Member State has failed to implement EU law properly the Commission will issue that State with a 'reasoned opinion' detailing the breach, and ask the Member State for its response. If this does not resolve the issue then the Commission will bring the matter before the ECJ under Article 226 EC Treaty. In practice the Commission is heavily dependent on the public and environmental NGOs for information about implementation failures. This system is poorly organised and Article 226 proceedings are initiated in only a small minority of cases (Krämer 1996). Grant (1996) comments that 'the

Commission's enforcement machinery has proven hopelessly inadequate and too open to political interference'. In order to reduce these inadequacies the Commission has encouraged the development of informal European monitoring networks such as the 'Network for the Implementation and Enforcement of Environmental Law' (IMPEL) (Hunter *et al.* 1998).

Fortunately, in those cases that do reach court, the ECJ is generally sympathetic to environmental interests (Sands 1996), as evidenced by the willingness of the court to utilise international environmental agreements in the interpretation of EU environmental law (Hey 1998).

Given the Commission's difficulties it is fortunate that environmentally concerned individuals or NGOs often make strenuous efforts to see that EU environmental law is enforced. They are, unfortunately, somewhat hamstrung in their efforts by legal rules which limit individuals' legal rights of standing to come before courts. The situation differs between the ECJ and national courts. Individuals and NGOs have very limited rights to come before the ECJ. The ECJ will only hear a person who is 'individually concerned' with the issue in question; that is, is more affected than the general public (*Stichting Greenpeace Council* v. *Commission*, Case C-321/95P). Since environmental laws affect populations fairly equally, it is uncommon for individuals to pass this test. For example, in *Danielsson* v. *Commission* [1996] an individual who might have been affected by the French Nuclear tests in the Pacific was denied 'standing' before the ECJ because

> that circumstance alone would not be sufficient to distinguish them individually . . . since damage of the kind they cite could affect, in the same way, any person residing in the area in question.

Krämer (1996) suggests revising the EU Treaty to include a citizen right of standing to challenge EU decisions and non-implementation of EU environmental law in the ECJ. Schikhof (1998) points out that this strict approach is in conflict with developments in national courts, and is inconsistent with the EU's own environmental policy.

Lack of rights before the ECJ leaves individuals with *national* courts as their best option for independent enforcement action. First an individual must convince the national court to grant him or her a right of hearing, known as 'standing' in legal parlance (see Chapter 5 for details). Having reached a national court the individual needs a substantive legal argument. Unfortunately EU law does not create a general legal right to complain, in national courts, of non-implementation or breach or EU law (*Rewe* v. *Hauptzollamt Kiel* [1981]). The best lines of argument are (1) that some national law has been wrongly interpreted so as not to give proper effect to its 'parent' EU law; (2) that the EU law generates rights for the individual that have been abrogated; or (3) that compensation is due for the failure to implement the EU law in question.

In the matter of interpretation national courts are bound to interpret their national law in keeping with EU law (*Marleasing SA* v. *La Commercial Internacional de Alimentacion SA* [1990]). In the second case, that of abrogation of rights, it is open to the individual to claim that the EU law that the Member State has failed to properly implement possesses *Direct Effect*. If a Directive is deemed to have Direct Effect then the courts must apply its terms, in favour of an individual, even where it has not been properly translated into national law. Unfortunately the ECJ has restricted the doctrine of Direct Effect to Directives or parts of Directives that are 'unconditional', 'sufficiently precise' and have the nature of creating rights for individuals. Environmental Directives which may have these characteristics include those providing drinking water standards (Directive 80/778), rights of access to environmental information (Directive 90/313), and public consultation during environmental impact assessment (Directive 97/11).

The practice of European national courts, whenever Direct Effect is claimed, is to refer the matter to the ECJ for its view. Unfortunately, the ECJ has shown little enthusiasm for applying Direct Effect in environmental cases. For example, in *Comitato di Coordinamento per la Defesa della Cava* v. *Regione Lombardia* [1994], the ECJ determined that the EU Waste Framework Directive, which requires waste to be disposed of 'without harm to man or the environment', does not have Direct Effect because Member States have discretion as to how to achieve its objectives.

Independently of the Direct Effect doctrine, individuals may be able to claim *compensation* for a Member State's failure to implement an EU environmental law properly. This right is limited to Directives which, by their nature, confer rights on individuals that are identifiable from the provisions of the directive, and where there is a *causal link* between the Member State's implementation failure and damage suffered by the person affected (*Francovich* v. *Italy* [1991]).

Member States may challenge the legality of each other's laws or actions under Article 227 EC Treaty (e.g. the French challenge of UK fishery conservation measures in *France* v. *United Kingdom* [1979]). Recourse to

Article 227 has been rare in the past but, given the recent trend for the imposition of unilateral trade sanctions by certain EU states, may become more prevalent in the future.

Problems in enforcing EU environmental law can be attributed to a variety of factors. EU environmental law lacks 'readily identifiable vested interests willing and able to secure enforcement'. There is an absence of direct economic impact. The Environment DG is small and relatively understaffed, relying largely on public complaints to bring Member States' failures in implementation to its attention. Environmental Directives are often worded imprecisely, allowing varying interpretations by Member States. Environmental Directives often ostensibly create discretions for Member States rather than requiring some specified action: it can be unclear whether failure to exercise such discretion amounts to an implementation failure. Finally, Article 226 proceedings are unlikely to ever result in full implementation of EU environmental law which requires, instead, a 'genuine *internal* political will' by Member States themselves (Macrory 1992).

European human rights law

An additional body of European law with significance for environmental protection is the European Convention of Human Rights supervised by the European Court of Human Rights in Strasbourg. The main provisions of the Convention with potential relevance to environmental interests are Article 2 (the right to life), Article 3 (protection against torture or inhuman and degrading treatment or punishment), Article 8 (respect for private life, home and correspondence) and Article I of the First Protocol (the right to property) (Desgagné 1995, 277).

Article 2 – the 'right to life' – imposes a duty on the State not to take away life. It might also require the State to take precautionary action against clear environmental risks threatening health or life (Desgagné 1995, 267). Article 8 – respect of privacy – grounded a successful claim in *Lopes Ostra* v. *Spain* (1994). The applicant lived close to a waste-water treatment plant authorised by the State, and suffered serious health problems and inconvenience as a result of the close proximity of the plant to her house. This was held to be an infringement of Article 8 of the ECHR.

Applications under the ECHR often fail because the Convention allows states to balance the rights involved against the need for economic development. For instance, aircraft noise has been held to be justifiable, even though it causes a serious deterioration of local environmental conditions (*Powell and Rayner* v. *UK* (1990)).

The ECHR has recently been incorporated into UK environmental law through the Human Rights Act 1998 and the Scotland Act 1998. This means that, in future, applicants will not have to undergo the lengthy delays involved in submission of their case before the Court in Strasbourg, but will be able to rely on the ECHR directly in UK courts.

Other regional environmental law

Europe is not the only source of regional law with relevance to environmental protection. Each major world region has some form of regional legal system – usually confined to matters of free trade but in some cases having a wider jurisdiction.

The 1994 North America Free Trade Agreement (NAFTA) between Canada, Mexico and the United States is destined to be a significant source of regional law with relevance to environmental protection. The NAFTA treaty has responded to concerns over 'downward pressure' created by free trade regimes in two ways: first, by encouraging harmonisation of standards and, second, by carefully elaborating the circumstances in which a party may restrict trade in order to protect the environment. Both techniques are applied to two areas of national law: sanitary and phytosanitary (S&P) measures and other non-S&P measures. Further protection is provided through restrictions on creation of 'pollution havens' where low environmental standards could apply.

S&P measures are measures 'to protect animal, plant or human life in a party's territory' from risks posed by pests, diseases, additives, contaminants, toxins or disease-causing organisms (Article 724). Each NAFTA party must harmonise its S&P measures with those of the other NAFTA party States 'to the greatest extent practicable . . . without reducing the level of protection of human, animal or plant life or health' (Article 714(1)). That said, NAFTA parties may, nevertheless, establish their own appropriate levels of protection for S&P measures, so long as this is done through a risk assessment taking into account, *amongst other things*, 'relevant scientific evidence' and 'relevant ecological and other environmental conditions' (Article 715(1)). In determining the 'appropriate level', and in conducting the risk assessment, economic impacts must be taken into account (Article 715(2)). NAFTA States must, in any event, 'avoid arbitrary or unjustifiable discrimination against the good of another Party or constitute a disguised restriction on trade between the Parties' (Article 715(3)(b)).

Non-S&P measures must be 'to the greatest extent compatible' to facilitate trade between the parties in goods or services (Article 906). Each party is permitted to establish 'the levels of protection that it considers appropriate' for 'legitimate objectives of safety or the protection of human, animal or plant life or health, the environment or consumers' (Article 904).

NAFTA's 'pollution haven' clause, Article 1114(2), admonishes the parties not to 'waive or otherwise derogate from . . . [environmental standards] as an encouragement for the establishment, acquisition, expansion or retention in its territory of an investor'. The weakness of this clause is its non-obligatory language.

Despite the features discussed above, NAFTA's inherent limitations in dealing with environmental standards, deriving from its nature as a trade agreement, led to the negotiation of the separate North American Agreement on Environmental Cooperation (NAAEC). The NAAEC's aims include protecting and improving the environment, promoting sustainable development, enhancing compliance with environmental laws and regulations, and promoting pollution prevention (Article 1). The main organ of the NAAEC, the North American Commission on Environmental Cooperation, may make recommendations on (for example) such matters as environmental monitoring, pollution prevention techniques, common indicators for reporting the state of the environment, and the use of economic instruments for environmental objectives. NAFTA's pollution haven clause, discussed above, is buttressed by the NAAEC's 'anti-roll back' clause requiring each party to 'ensure that its laws and regulations provide for high levels of environmental protection and shall strive to continue and improve those laws and regulations'. Finally, a unique feature of the NAAEC is the provision enabling each party to ensure adequate enforcement of environmental standards by the other NAAEC parties.

Africa has several regional organisations that facilitate the creation of regional laws (Sands 1995, 715). Although these are largely oriented towards economics and development, some also have a role in environmental protection. In 1991 the 50 members of the Organisation for African Unity (OAU) signed the Abuja Treaty, creating the African

Economic Community (AEC) (Thompson 1993) which aims to achieve a step-wise evolution to full economic integration between African states by 2020.

The Abuja Treaty is predominantly concerned with development, trade and political cooperation rather than environmental protection. However the environment is not totally ignored. Article 3, which sets out the Community's guiding principles, refers to the need for 'promotion of a peaceful environment as a prerequisite for economic development'. The party states undertake to 'promote a healthy environment' (Article 58). Article 74 calls on African countries to adopt national population policies and mechanisms to ensure a balance between population growth and socio-economic development.

In Asia the main regional bodies are the Association of South East Asian Nations (ASEAN) and the Asia Pacific Economic Cooperation Forum (APEC). ASEAN was formed in 1967 by Indonesia, Malaysia, the Philippines, Singapore and Thailand to promote political and economic cooperation and regional stability. Since then Brunei, Vietnam, Laos and Burma have entered the Association. In 1992 the heads of government decided to establish an ASEAN Free Trade Area over a fifteen-year period.

Although not vested with legislative powers, ASEAN commands influence on Asia–Pacific trade, political, and security issues, largely through its use of consultation and cooperation. The ASEAN Ministerial Meeting on Environment (AMME) meets every three years to discuss progress in environmental protection. AMME has adopted the Manila Declaration and the Bangkok Declaration: the former setting out policy guidelines on the environment; the latter noting progress and adopting guidelines in specific environmental sectors (Rose 1995). In 1990 ASEAN ministers agreed the 'Kuala Lumpur Accord' (Anon. 1991), giving priority to environmental management and setting out expectations in relation to matters such as the harmonisation of environment standards and transboundary pollution prevention.

APEC was formed in 1969 to advance and facilitate the economic integration of a broad span of Asian states and their trading partners. APEC now includes Australia, Brunei, Canada, Japan, Hong Kong, Indonesia, Malaysia, Mexico, New Zealand, Papua New Guinea, China, The Philippines, The Republic of Korea, Singapore, Taiwan, Thailand and the United States. A limited amount of joint action has been taken by APEC states in relation to the environment, including, in 1995, agreement of a set of principles to guide environmental activities in APEC's working groups and committees (Anon. 1995).

A range of other regional bodies with law-making or facilitating functions exist in Asia: e.g. the UN Economic and Social Commission for Asia and the Pacific (ESCAP); the Regional Office for Asia and the Pacific of the UN Environment Programme (ROAP); the International Maritime Organisation (IMO) (for details see Rose 1995).

National environmental law

Despite the huge growth in international and regional laws relating to environmental management, national law is still pre-eminent. In a world system of over 170 states 'domestic' law has the greatest impact on and control over the behaviour of individuals. There are several reasons why national law is critical in environmental regimes. First, national governments remain in a strong position to pass effective environmental legislation and have every incentive to do so due to the geographic coexistence of the legal regime and the environmental region in question. Second, international and regional law, as we have seen, frequently requires to be translated into domestic level law for legal effect and almost always requires to be enforced at that level. National law is the 'sharp end' of regional or global environmental policy. Third, national law may be among the most effective methods for giving effect to the 'environmental rights' of individuals, animals and plants or the environment in general (Birnie and Boyle 1992, 188). Finally, national law may serve to shape and influence the content of regional and international laws. The process is not all 'top-down'; many significant principles and standards begin with one or a small number of states and spread until they are embedded in a broader legal context.

Creating national environmental legislation

National environmental legislation is the outcome, in western democratic states, of the processes of a legislature or Parliament. (States with religious legal systems or in which power is taken by military coups tend to be exceptions.) Although the details of parliamentary process differ between states (Blondel 1973; Olson 1980), certain common features exist. Parliaments are normally (but not always) *bicameral* (i.e. consist of two chambers). There may also be an element of monarchical or presidential power. In the UK, constitutionally, 'the Queen in Parliament' enacts legislation. In practice, given strong parliamentary majorities, governments are able to 'force through' even unpopular measures. It is arguable that, because of this effect, the legislative process is undemocratic (i.e. that we have a system of temporary elected dictatorships). As far as environmental legislation is concerned this may be no bad thing since behavioural changes required to protect the environment are not always going to be popular at the time of their creation.

Usually, lower houses consist of elected representatives (Mill 1958) who are expected, subject to the constraints of the party system, to apply their own judgment and intelligence to matters at hand rather than act as mere delegates for popular sentiment. This feature provides a partial safeguard from the problem of 'tyranny of the majority' which besets direct forms of democracy. Representative democratic theory leads us to expect intelligent educated politicians to make sensible responses to environmental problems. This raises a collateral question of whether Members of Parliament, engaged in debating and legislating for environmental matters, should be required to have had some formal education in environmental studies and environmental law.

The function of the upper house is usually taken to be scrutiny of legislation proposed by the lower house, and restraint of excessive or overly populist legislative proposals. Upper houses may be constituted by birth, nomination or election. Election, in such cases, is often on the basis of proportional representation. In the UK the tradition of lords based on hereditary title was abolished by the House of Lords Act 1999. Currently the lords are constituted by nomination for life ('life peers') with a small residual number of hereditary peers.

In most nations parliamentary power is subject to constitutional limits. In the United States, for example, the President can veto Congressional enactments and Congress must approve by a two-thirds majority any bill it wishes to pass despite a presidential veto. The first ten amendments to the US Constitution (i.e. the Bill of Rights), and subsequent amendments, limit federal and state governmental powers. The Supreme Court is empowered to determine whether government legislation breaches these fundamental rights and, if so, to strike it down as unconstitutional.

In the UK, the traditional constitutional position is that the sovereignty of Parliament is subject to no limits (Dicey 1965, 39). In reality Parliament's legislative powers have for some time been limited by the obligation to comply with European Union law (Art. 10 EC Treaty) and the European Convention on Human Rights (ECHR). In relation to the latter the Human Rights Act 1998 provides that, so far as it is possible to do so, the courts must read and give effect to UK legislation in a way which is compatible with ECHR right. In cases where this is not possible a court may make a 'declaration of incompatibility' – intended to prompt Parliament to bring UK law into alignment with the convention.

The legislative process

In most states, legislation is proposed by the government of the day, debated by the legislature, subject to scrutiny, passed to an upper chamber, and then formally ratified. The following discussion uses the UK to exemplify the process (Miers and Page 1990; Zander 1980; c.f. for the US system, Kubasek and Silverman 2000). UK Acts of Parliament usually derive from government policy. This will either be pre-existing policy, arising from election commitments to legislate on certain topics (e.g. the Transport bill 1999), or ongoing policy due to some pressing but unforeseen issue (e.g. the Bovine Spongiform Encephalopathy Order 1991 arising from the discovery of BSE in cattle). In the UK, environmental law proposals are also stimulated by reports of a permanent law/policy review body – the Royal Commission on Environmental Pollution (RCEP).

Often the detail of the proposal follows public consultations in the form of a 'green' or 'white' paper (*Hansard* 1992, 403). Environmental groups have no statutory rights to be consulted on proposed environmental laws but, in practice, the larger, more respectable environmental NGOs are given a hearing by the relevant government department. Pro-business interest groups will also lobby and seek government consultation. In some nations – e.g. the USA – there is a culture of payments or 'contributions' as a component of lobbying. The extent to which lobbying is successful in determining legislative outcomes is not clear. Business-oriented lobbying in the USA, for example, vastly outweighs that of environmental groups. According to Goldsmith (1993, xii) 'the lobbying efforts of powerful industrial groups intent on defending their petty short-term interests, come what may [are] a major constraint on government action to tackle the serious environmental problems that face us'. Vastly more is spent on lobbying by anti-environmental interests than by those who support environmental legislation. Yet environmental regulation shows steady growth. It is possible that this apparent paradox is explicable by reference to the inability of industry to lobby coherently, and the possibility that businesses see advantages to environmental regulation (Wilkinson 1997).

Occasionally environmental NGOs initiate environmental laws by persuading an MP to take up their cause and introduce legislation as a Private Members' bill. In the UK Parliament bills are not limited to those introduced by government ('public bills'): a limited amount of parliamentary time is allocated to the introduction of bills by individual MPs (Private Members' bills). A Private Members' bill which does not receive cross-party support can usually be defeated by the tabling, and discussing at length, of amendments, thus preventing the passage of all the required parliamentary stages in the allotted time. This is referred to as 'talking a bill out'. The government, who determine the legislative timetable, can easily rescue a Private Members' bill by giving up time allocated to government business but, due to the pressing legislative agenda, governments are often reluctant to do this. Consequently, only around 20 per cent of Private Members' bills make it onto the statute book (see HMSO 1999a, 1999b). That has not prevented several notable successes, including the Road Traffic Reduction (National Targets) Act 1998.

When the government is convinced of the need for legislation, a 'bill team' is assembled within the relevant department (usually, for UK environmental law, either the Department of Transport, Local Government and the Regions (DTLR) or the Department for Environment, Food and Rural Affairs (DEFRA)). The bill team lawyers, having settled upon the content of legislation, and having obtained approval of the Cabinet's Future Legislation Committee, instruct the Office of Parliamentary Counsel who draft public bills. Once Parliamentary Counsel have concretised the proposal in written form, the bill is given a first reading in the House of origin (either the House of Commons or the House of Lords, but usually the former). After this – which amounts to no more than reading out the title of the bill – the bill is printed and made available to the public (see HMSO 1999c). The fact that it is not available until *after* the first reading is arguably a deficit in public consultation and governmental transparency (*Hansard* 1992). The public are often unaware of forthcoming legislation and, therefore, unable to voice their views on the substance of the proposal to their MP.

The bill then moves to its second reading in the same House, at which point vigorous debate of its substantive content takes place on the floor of the House. At this stage, public attention may be drawn to the bill through media coverage, and environmental pressure groups begin or increase their vociferous attention. The debates are recorded in a public record (*Hansard*) which is available in libraries and on the internet (http://www.parliament.the-stationery-office.co.uk/pa/cm/cmhansrd.htm). Although this debate is theoretically public, much of the detail of legislation is agreed though pre-bill negotiations (Harris 1993, 144). Parliament receives 'what is, to all intents and purposes, a finished product' (Mackintosh 1982, 144).

The bill next goes before a committee ('committee stage') which scrutinises a selection of the proposed amendments to and clauses of the bill. In the case of environmental legislation the relevant committee is a permanent or '*standing*' committee which allows its members to build up expertise in environmental issues. The committee reports to the House and the bill moves forward with such amendments as are agreed to.

After its third reading in the House of origin, the bill passes to the other House where it must pass through the same stages again. The committee stage in the Lords takes place on the floor of the house – providing greater openness. The House of Lords can attempt to force the House of Commons to 'think again' by defeating bills or making amendments that would not be acceptable to the Commons. But if the Lords persist, the Commons can force the bill through by virtue of the Parliament Acts of 1911 and 1949 (e.g. the War Crimes Act 1991).

After its third reading in the second House the bill receives, as a formality, the Royal Assent and becomes an Act of Parliament (i.e. law). Shortly after Royal Assent it is published by HMSO in hard copy and, since 1997, on the Internet (http://www.hmso.gov.uk/acts.htm). This is not, however, the end of the matter. Two further steps are often required to give legal effect to the law concerned: 'coming into force', and the production of 'delegated legislation'.

Legislation does not usually take effect for the future immediately after Royal Assent. Each section of each Act is required to 'come into force'. This may happen automatically, because the Act itself specifies a date or dates for its sections to take effect. Frequently, however, the section of an Act will require a further triggering instrument to come into force. This 'trigger' usually takes the form of an Order of the Queen in Council. This built-in delay can have both desirable and undesirable consequences for environmental protection. Positively, it provides a further opportunity to allow businesses to alter practices in an environmentally protective manner. The existence of delay may, however, mean that sections of statute are not brought into force in good time or, in certain cases, not at all. The implementation of the Control of Pollution Act 1974, for instance, was notoriously slow (Evans 1992, 156). The 'in force' process also creates a good deal of uncertainty for businesses and ordinary members of the public in knowing which laws they must comply with. It is possible to ascertain which sections of an Act have come into force by referring to the publications *Is it in Force?* and *Current Law*, or by reference to LEXIS, but these are not widely available.

An Act of Parliament can only rarely, and in the most simple cases, provide for every contingency that may arise, and provide the full detail of a regulatory regime (although the complexity of some statutes indicates that this point is sometimes missed by Parliamentary draftsman). Consequently much law, and much environmental law in particular, takes the form of 'delegated' or 'secondary' legislation. The normal form of delegated legislation is *regulations* made in a form known as a 'statutory instrument'. Environmentally relevant regulations are generally drawn up by the Secretary of State for the Environment. Once drawn up they are then subject to a form of parliamentary approval. Depending upon what the 'parent' Act specifies this is either a *negative resolution* or an *affirmative resolution* procedure.

Delegated legislation relieves pressure on Parliamentary time and provides a fast and flexible mechanism for creating the 'fine print' of a system of legislative control. A usual criticism of delegated legislation is that the 'devil is in the detail' and that the detail, in such measures, is not subject to full democratic process. This results in delegated legislation being more technocratic and expert-led, which may or may not be to the advantage of the environment.

The quality of UK legislation, including environmental statutes, is not always as high as it might be. The Renton report (HMSO 1975) highlighted the linguistic complexity and difficulty of comprehension of the end product. As Jane Hern, a member of the Law Society, put it:

> Lawyers are constantly trying to find out what [statute law] is in force, when it came into force . . . and when they have found out, what the hell it means.
>
> (Hern 1992, cited by The Hansard Society 1992, 1)

This complexity and density is partly explicable by reference to the very literal method of interpretation adopted by the courts:

> It is because the judges have not felt it right to fill in the gaps and have been giving a literal interpretation for many years that the draftsman has felt that he has to try and think of every conceivable thing and put it in so far as he can so that even the person unwilling to understand will follow it.
>
> (HMSO 1975)

Some help in understanding environmental legislation is provided in the form of government guidance. Many environmental statutes provide legal authority for such guidance. Usually, such guidance should be taken into account by courts, government ministers, etc., but is not strictly binding (*Tesco Stores Ltd.* v. *Secretary of State* [1995]). Occasionally, the guidance itself has legal effect as in the case of the guidance concerning the statutory regime for contaminated land under Section 57 of the Environment Act 1995.

A difficulty with the use of national legislation for environmental protection is that environmental concerns may be overlooked in legislation originating in non-environment departments. Integrating environmental considerations into all areas of law poses considerable challenge to governments and is arguably now the single most important issue in the drive for sustainability (cf. Sumikura 1998). Even environmental legislation may not be as protective of nature as we would expect. It is often high in rhetoric, low in impact. We should be prepared to consider that much environmental law may be created to fulfil symbolic political functions, rather than to radically alter human–nature relations (Aubert 1966; Wilkinson 1997). As Stewart (1977) observes in relation to US environmental laws:

> The statutory objectives and deadlines were in many cases unrealistically ambitious, perhaps intentionally so. Sponsors of the 1970 Clean Air Amendments and the 1972 FWPCA may well have deliberately selected radical goals and unrealistic deadlines – such as the elimination of all discharges of pollution into navigable waters by 1985 – in order to dramatize environmental issues, arouse public support for legislation, and persuade polluters that the federal government was committed to substantial control measures.
>
> (Stewart 1977, 1199)

Finally, we must question whether legislatures, and the laws that they produce, can ever be radically green since, as Dobson argues, part of the participatory programme that green thought demands involves a shift away from the traditional centralised law-making processes (Dobson 1990, 132).

In addition to legislation, at least in common law states, national environmental law comprises judge-made law through the doctrine of precedent (see Chapter 1). The areas of common law most directly relevant to environmental protection are the laws of nuisance, trespass, the so-called 'rule in *Rylands* v. *Fletcher*', and negligence.

Nuisance gives those with an interest in land a remedy (monetary compensation or an injunction) for an unreasonable interference with the *use* of land. Nuisance has the advantage of flexibility: the notion of 'unreasonable interference' is situation-dependent and can adapt quickly to cover novel types of pollutant or environmental damage. However, as a mechanism for environmental protection, nuisance has several weaknesses including, notably, a requirement that environmental damage be 'reasonably foreseeable' and the so-called 'locality doctrine' which allows for *higher* levels of interference or pollution in industrial or pre-polluted areas development (*St Helens Smelting* v. *Tipping* (1865)). Damages cannot be claimed in the UK using nuisance for environmental damage that is not 'sensible'; that is, visible to the human senses (*Salvin* v. *North Brancepeth Coal Co.* [1874]). Thus the law will not compensate for contamination that can only be shown by scientific inquiry.

The action for trespass is available whenever there has been a *direct* interference with land (or, by analogy, air or water). So, for instance, an action could be brought in trespass where waste or chemicals had escaped from land onto a neighbouring property. Such a state of affairs often results in interference with the *use* of land: hence the availability of the action in trespass, in practice, often overlaps with the availability of nuisance. Trespass has the advantage that, at least when an injunction is sought, it is actionable *per se* (i.e. no actual environmental damage need be proven). A distinct disadvantage of trespass is the requirement that the invasion be *direct*, which will often rule out the use of trespass in cases of pollution where the pollutants have travelled by unpredictable means to their eventual resting place. In *Esso Petroleum Ltd.* v. *Southport Corporation* [1954] oil was jettisoned at sea in order to save a ship that had run aground. An action in trespass for the pollution of the corporation's nearby beach failed: the oil had been 'committed to the action of the wind and wave, with no certainty, so far as appears, how, when or under what conditions it might come to shore' (i.e. the pollution was not 'direct').

In the nineteenth century, in *Rylands* v. *Fletcher* (1868), the House of Lords stated that:

> the person who for his own purposes brings on his lands and collects and keeps there anything likely to do mischief if it escapes, must keep it in at his peril, and, if he does not do so, is prima facie answerable for all the damage which is the natural consequence of its escape.

Commentators and judges alike assumed this to have created a new rule of strict liability for escaping dangerous things. The lack of any requirement to show unreasonable use, or lack of care, gives *Rylands* v. *Fletcher* an advantage over nuisance or trespass. Unfortunately *Rickards* v. *Lothian* [1912] restricts the applicability of *Rylands* to damage resulting from '*non-natural*' uses of land. It is unclear whether 'non-natural' refers to things not commonplace or ordinarily on land or, more broadly, to things that are of human origin (e.g. chemicals). One theory that has prospered from time to time is that activities that bring benefit to the local community cannot count as non-natural, although this view was recently rejected by the House of Lords in *Cambridge Water Company* v. *Eastern Counties Leather* [1994].

To succeed in negligence a person must prove (a) the existence of a duty of care, (b) a breach of that duty leading to (c) damage that is (d) caused by that breach. Each of these four stages can be problematic. Nevertheless, negligence has potential as a mechanism for providing compensation in cases of environmental damage. For instance, in *Swan Fisheries Ltd* v. *Holberton* (1987) a farmer who blocked off a stream in order to refill a dam was liable in negligence for the death of fish downstream. The farmer has a duty to consider downstream users of the river. As he had not done so, and as a consequence environmental damage was caused, he was liable to pay compensation. The big advantage of negligence is that it can theoretically apply between any type of actors: it is not limited to actions between neighbouring landowners. Thus in *Scott-Whitehead* v. *National Coal Board and Southern Water Authority* the authority, which issued permits for discharges into and abstractions out of a river, was liable in negligence for failing to warn a farmer, who relied on river water for irrigation purposes, about inputs of saline water that it had authorised from a mine owned by the coal board. More dramatically, in the litigation that ensued after the catastrophic *Amoco Cadiz*, the cargo owner's insurer, Petroleum Insurance Limited, was awarded the equivalent of £21.2 million against Amoco Corporation and Astilleros (Spanish designers and builders of the *Amoco Cadiz*) for the loss of Royal Dutch Shell's cargo of oil. Companies that fail to exercise due care in the design and construction of pollution control technology may find themselves liable in negligence to those affected by any ensuing pollution.

2

...mon law (in) action: *the Cambridge Water ...ompany Case* [1994]

The Cambridge Water Company case provides an interesting example of the failure of the common law to protect the environment. The litigation arose from chronic pollution of an aquifer (a natural underground reservoir of water held in saturated bedrock) by a tanning company, Eastern Counties Leather (ECL). The Cambridge Water Company (CWC) relied upon the aquifer to satisfy its statutory obligation to maintain a drinking water supply to the Cambridge area. In the 1950s, ECL, who were located in a village near to the aquifer, began to use the chemicals Trichloroethene (TCE) and Perchloroethene (PCE) as degreasing agents in the tanning process. Whilst moving the chemicals around their site they routinely spilled small amounts. These eventually penetrated the soil and migrated slowly down into the aquifer.

In the 1980s, the European Community and the World Health Organisation (WHO) set health-based standards permitting only minute quantities of TCE and PCE in drinking water.

In 1983 CWC became aware that water abstracted from the aquifer was contaminated by TCE and PCE and could not be supplied for public consumption. In consequence, CWC shut down the bore hole that abstracted the polluted water and moved their pumping operations to a part of the aquifer not affected by the contamination. The company then sought damages of £900,000 from ECL, basing their claims on three 'heads' of law: (a) nuisance, (b) negligence, and (c) the so-called 'rule in *Rylands* v. *Fletcher* (1868), L.R. 3 H.L. 330.

The claims in nuisance and negligence failed because the water company could not overcome the obstacle created by an earlier precedent; that is, that liability in nuisance and negligence extends only to *reasonably foreseeable* damage (i.e. damage foreseeable by a reasonable person standing in the defendant's shoes at the relevant time). The trial judge Ian Kennedy J found that, on the facts, the reasonable tannery supervisor, alive to the pattern of spillage that had occurred, would not have foreseen any resulting environmental hazard.

The claim based on *Rylands* v. *Fletcher* also failed. The trial judge felt that a tannery – an activity providing employment to the local area – was not a 'non-natural' use of land. It was unclear whether the liability under *Rylands* was limited to foreseeable damage. When the litigation case reached the House of Lords this position was reversed. Contrary to the trial judge, their Lordships concluded that the storage of substantial quantities of chemicals on industrial premises was to be regarded as 'an almost classical example of non-natural use'. They also considered the relationship between nuisance and *Rylands* and concluded that liability in the latter was, as in the former, limited to reasonably foreseeable damage. Since the 'environmental hazard' created by the leaching chemicals was not reasonably foreseeable, liability did not lie.

This case was, on balance, a bad result for the environment. Admittedly, it did push the concept of 'non-natural use' in the right direction. But the imposition of a foreseeability test on strict liability was not a positive move. Given the nature of science and the gradual evolution of knowledge, industrial processes often trigger pollution or environmental damage of a type or extent not easily predictable at the outset. One only has to consider the history of issues such as BSE, nuclear waste, and ozone-depleting substances to realise that on each occasion substances or processes that are initially thought to be safe later turn out to have unpredictable (or at least unpredicted) negative environmental side-effects. The decision in *Cambridge Water* makes it virtually impossible to impose future liability on those who cause these environmental hazards. This, in turn, reduces the deterrent effect on those now contemplating novel and potentially hazardous activities. It thus pulls in the opposite direction from the precautionary principle (see Chapter 4).

More worrying is the judicial apathy that the case reveals towards the English common law and its overall role in environmental protection. Lord Goff, giving a unanimous opinion of the court, stated:

> The protection and preservation of the environment is now perceived as being of crucial importance to the future of mankind; and public bodies both national and international are taking significant steps towards the establishment of legislation that will promote the protection of the environment.. But it does not follow from these developments that a common law principle should be developed or rendered more strict. On the contrary, given that so much well informed and well structured legislation is being put in place for the purpose, there is less need for the courts to develop a common law principle to achieve the same end and indeed it may well be undesirable that they should do so.

In other words, the common law will not reach out its protective mantle because environmental protection is seen as a matter for European and UK legislatures. That is a dangerous abdication of the English common law's moral guardianship of nature. It fails to recognise that, as the facts of the case show, not all environmental harm will be caught by environmental legislation. Despite Lord Goff's reference to 'significant steps' there is, as yet, no comprehensive law imposing liability on those who harm the environment. Common law courts should have the courage to develop their own protective principles until that day arrives.

Viewed 'in the round', the common law's delineation of property rights, coupled with clear and enforceable liability rules (nuisance actions), may provide a better balance between economic and ecological interests than that provided by the statutory systems of regulatory control that have become prevalent in modern times (Yandle 1997). On the other hand, the range of restrictions to each class of action, the risk of legal costs and the requirements in most cases for a legal interest in land, together indicate the need for a strong statutory framework in addition to common law protection.

Enforcing national environmental law

Discussion of the various techniques used to implement and enforce national environmental law is reserved until Chapter 5.

Conclusion

As we have seen, many factors influence the formation and effectiveness of environmental law, at whichever level it is produced. The important thing, from the environmental perspective, is to ensure a high degree of correspondence or fit between law at each level. It is also important to ensure that the legal source fits the problem; that is, that international law is used for tackling problems of global dimensions, such as global warming, and that national law is used for preserving nationally important aspects of the environment.

Key points

- International law provides an important framework for the resolution of certain environmental issues. It is, however, affected by problems, including its essentially voluntary nature, lack of certainty and lack of effective enforcement mechanisms.
- Regional law has the potential to provide stronger and more harmonised environmental standards and to combine environmental protection with economic development. EU law is a major driving force of national European environmental law but suffers from problems of patchy enforcement and lack of public access to justice. In America, NAFTA is set to become an increasingly important framework for the setting of environmental standards.

- National law is vital in the fight against environmental harm. It is as good, but only as good, as the legislative process and judicially created rules from which it derives. Creative legislatures and activist judiciaries are important prerequisites of strong national laws for environmental protection.

Further reading

Boyle, A.E. (1991) 'Saving the World? Implementation and Enforcement of International Environmental Law through International Institutions', *Journal of Environmental Law*, vol. 3, no. 2, pp. 229–45. This article, although somewhat dated now, provides an excellent antidote to the view that the lack of authoritative enforcement mechanisms renders international law impotent. The author focuses on the critical role played by international institutions in administering, monitoring and in other ways securing compliance with IEL.

Schoenbaum, T.J. (1997) 'International Trade and Protection of the Environment: The Continuing Search for Reconciliation', *American Journal of International Law*, vol. 91, pp. 268–308. This paper examines the tension between promotion of international free trade regimes and sovereign authority to protect nature.

Discussion questions

1. Is it justifiable for states to be able to choose whether or not to be parties to international environmental conventions? If not, what is the solution to this problem?
2. Is any one source of environmental law particularly well equipped to deal with environmental degradation?
3. Should regional law, such as the law of the European Union, be seen as a threat to national efforts to protect the environment, or as a main pillar of environmental protection?
4. Consider ways in which one of the main legal sources of environmental protection could be improved.

Principles of environmental law

- **The nature of legal principles**
- **The preventative principle**
- **The precautionary principle**
- **The polluter pays principle**

Protection of the environment is, increasingly, seen as a question of 'principle'. Frequently, one reads or hears references to phrases such as 'the precautionary principle' or the 'polluter pays principle'. Often, in the context of developed versus developing states references are made to principles of 'environmental justice' and 'sustainable development'. This chapter explores the role of legal principles in relation to environmental protection, and examines in further detail individual environmental principles such as the precautionary principle.

The nature of legal principles

A principle is 'a source of action' or 'a general law or rule adopted or professed as a guide to action' (*Oxford English Dictionary*). It differs in three ways (Perry 1997) from a rule:

1. Rules operate in an all or nothing fashion. If a case falls within the conditions of application for a rule, then the rule must be applied (i.e. it is conclusive). Principles, by contrast, have the quality of 'weight' or importance.
2. Rules specify a course of action to be followed or avoided. Principles, by contrast, are about *values* and cannot, therefore, indicate particular actions.
3. Rules can be justified by principles, but the reverse is not true (Moore 1997).

Cheng (1953, 376) makes a similar point in relation to international law: a *rule* of law is a 'practical formulation of the principles' and the 'application of the principle to the infinitely varying circumstances of practical life [which] aims at bringing about substantive justice in every case'.

The great advantage of legal principles is that, given their general or non-particularised nature, they have the flexibility to deal with novel and complex situations, and to act as general guides to action. Yet this flexibility implies some unpredictability. Sometimes principles are seen as possessing unquantifiable legal obligations. For this reason developed countries have not always supported the inclusion of environmental principles in treaties and other legislation. The US, in particular, was opposed to the inclusion of general environmental principles in the 1992 Climate Change Convention, fearing the creation of obligations and commitments of uncertain dimensions. On the other hand states sometimes perceive principles as weaker than (hence preferable to) strict rules or standards. The UK, also, made a declaration upon signature of the 1992 Biological Diversity Convention stating that the obligation in Article 3 of that convention – prohibiting transboundary harm – was one of principle, implying that this created no firm legal obligation (Sands 1995, 185).

The criticism that principles are vague and indeterminate misses the point that principles are, by definition, *general* guides to action; they do not, and are not intended to, provide specific rules of behaviour or precise technical standards.

Legal principles are found in both the 'unwritten' pronouncements of the judiciary (Van Hoecke 1995) and written law (i.e. statutes, treaties, etc.). The status held by unwritten principles in legal theory is disputed. Raz (1990) maintains that unwritten principles serve a *rationalisation* function (i.e. they justify a choice of firm legal rules). Perry (1997) argues, to the contrary, that legal rules are, or should be, a *summary* of the balance of underlying principles (i.e. that principles have priority and 'lead the way'). The difference between the two models arises in those crucial cases in which a court finds that a rule and its underlying principles are divergent. In the 'rationalisation' model the principles should be modified to account for the difference. In the 'summary' model the reverse is true.

Principles – whether written or unwritten – can also be divided into three categories (Alder and Wilkinson 1999): legally substantiated principles, legally binding principles and guiding principles. A *legally substantiated* principle is a principle that originates external to law (e.g. as a principle of environmental *policy*) but which is manifest and concretised through specific legal examples. As we shall see below, a case in point is that there are now enough legal instances of the polluter pays principle to say, with some confidence, that this is a legally substantiated principle.

A legally *binding* principle is one that must, by law, be followed by certain actors: in international law, states; in regional law, regional bodies; and in national law, government ministers and public bodies. By way of example, an established principle of administrative law is that, when exercising discretion, public bodies and ministers must act within the powers that they have been granted by Parliament. This entails the *reasonable* exercise of discretion (*Associated Picture Houses Ltd.* v. *Wednesbury Corporation* [1948]); that is, avoiding decisions that no sensible person could have reached (*Council of Civil Service Unions* v. *Minister for the Civil Service* [1985]).

A *guiding* principle is one that must be taken into account in the formulating or subsequent interpretation of legislation. Thus Article 3 of the 1992 Climate Change Convention requires party states, in achieving the objectives of the convention, to be guided by five principles:

> 1 The Parties should protect the climate system for the benefit of present and future generations of humankind, on the basis of equity and in accordance with their common but differentiated responsibilities and respective capabilities . . .
> 2 The specific needs and special circumstances of developing country Parties, especially those that are particularly vulnerable to the adverse effects of climate change . . . should be given full consideration.
> 3 The Parties should take precautionary measures to anticipate, prevent or minimize the causes of climate change and mitigate its adverse effects . . .
> 4 The Parties have a right to, and should, promote sustainable development . . .
> 5 The Parties should cooperate to promote a supportive and open international economic system that would lead to sustainable economic growth and development in all Parties, particularly developing country Parties . . .

Article 174 of the EC Treaty also contains guiding principles. This states that:

> Community policy on the environment shall aim at a high level of protection taking into account the diversity of situations in the various

> regions of the Community. It shall be based on the precautionary principle and on the principles that preventive action should be taken, that environmental damage should as a priority be rectified at source and that the polluter should pay.

Legislation can be rejected or declared unlawful in the event that relevant guiding principles, such as those in Article 174, are not taken into account in its formulation. Since Article 174 serves to define the general environmental objectives of the Community (*Peralta* [1994]) the Council retains responsibility for deciding upon specific action. But the Council does not have unlimited discretion. Specifically, if the Council commits 'a manifest error of appraisal regarding the conditions for the application' of Article 174 then, despite earlier contrary views (Krämer 2000), the ECJ has shown that it is prepared to judge the validity of the environmental legislation concerned (Doherty 1999). In *Safety Hi-Tech* v. *S & T, Gianni Bettati* [1999], for example, the ECJ was asked whether an EU Regulation for protection of the ozone layer, which prohibited the use of HCFCs in fire-fighting equipment, was consistent with the Article 174 which requires that 'Community policy on the environment shall aim at a high level of protection'. Safety Hi-Tech alleged that, in adopting a Regulation that excluded consideration of the global warming potential of a broad range of substances, the Council had not given priority to the highest level of environmental protection. The ECJ's answer was that the EU does not have to attain the *highest* possible protection: it may legislate to protect specific environmental entities such as the ozone layer. Thus, the Council, in passing the Regulation, had not committed a 'manifest error of appraisal'. The importance of the case lies in the demonstration by the ECJ judiciary of their willingness to utilise the environmental principles of the EC Treaty as the basis for legislative review (Doherty 2000).

Specific principles of environmental law

In an ideal world there might be some document or legal instrument containing a transparent 'master list' of all relevant environmental principles. The nearest to such a blueprint is the ambitious set of twenty-two 'proposed legal principles for environmental protection and sustainable development' adopted by the Expert Group on Environmental Law of the World Commission on Environment and Development (WCED 1987). These principles were grouped into general principles, rights, and responsibilities; principles, rights and obligations concerning transboundary natural resources and environmental interferences; and principles for state responsibility and the peaceful settlement of disputes. Unfortunately, although they originated in the highly influential 'Brundtland Report', no systematic attempt has followed to transform the proposed principles into law.

In order to ascertain which principles exist and the content of those principles it is necessary to look to the law itself. On inspection, the following list of principles has a fairly high degree of consensus:

- the preventative principle;
- the precautionary principle;
- the polluter pays principle;
- the proximity principle;
- the principle of sustainable development.

Of these, sustainable development operates as a meta-principle, under which the others are organised and to which they contribute. This chapter focuses on the first three, being important at every level of environmental law.

The preventative principle

The preventative principle is the notion that states, corporations, or individuals have, in certain circumstances, an obligation to take steps to avoid causing certain types of environmental damage to the environment, including the environment beyond their own territory or property ownership.

At the international level the preventative principle stems from a number of decisions of *ad hoc* tribunals. In *The Trail Smelter Arbitration* [1941] – a case arising from crop damage to property in the United States resulting from emissions from a Canadian smelter – the tribunal concluded that

> under the principles of international law, as well as the law of the United States, no State has the right to use or permit the use of its territory in such a manner as to cause injury by fumes in or to the territory of another or the properties or persons therein, when the case is of serious injury and the injury is established by clear and convincing evidence.

The *Corfu Channel* case (*UK* v. *Albania* (1949)) added to this, by establishing that every state is under an obligation not knowingly to allow its territory to be used for acts contrary to the rights of other states. These early propositions were useful but lacked an essential element: reference to those areas of the environment which lie outside of the territorial extent of any individual state (e.g. the high seas). This deficit was partially remedied by Principle 21 of the 1971 Stockholm Declaration on the Human Environment:

> States have, in accordance with the Charter of the United Nations and the principles of international law, the sovereign right to exploit their own resources pursuant to their own environmental policies, and the responsibility to ensure that activities within their own jurisdiction or control do not cause damage to the environment of other States *or of areas beyond the limits of national jurisprudence* [emphasis added].

Principle 21 was reiterated, almost unchanged, in the 1992 Rio Declaration on the Environment and Development and has been highly influential, now being embodied in a great many environmental conventions (Sands 1995, 196). Of these Article 194(2) of the United Nations Convention on the Law of the Sea, is fairly typical:

> States shall take, individually or jointly as appropriate, all measures consistent with this Convention that are necessary to prevent, reduce and control pollution of the marine environment from any source, using for this purpose the best practicable means at their disposal . . .

Although both the *Trail Smelter Arbitration* and Stockholm Principle 21 appear, on the surface, to amount to an absolute prohibition of extraterritorial environmental damage (Birnie and Boyle 1992, 94) other factors dilute that obligation. First, the obligation is imprecise and cannot easily be converted into concrete standards. In the US case *Amlon Metals* v. *FMC Corp., Inc.* (1991) the court observed that Principle 21 does not set forth any specific proscriptions, but rather refers only in a general sense to the responsibilities of nations not to cause damage to extraterritorial environment.

Second, the preventative obligation is probably merely an obligation to prevent injury to human health, or to prevent property/environmental damage, rather than a duty to prevent all contamination. This implies the acceptability of low-level pollution. A third and related point is that the principle must be weighed against a further principle: namely 'equitable utilisation' of resources, including resources such as pollution 'sinks' in the form of oceans and rivers. Fourth, concrete expressions of the preventative principle often limit the obligations of states to that of utilising the 'best available technology' to avoid pollution. Such expressions are commonly formulated to take account of the lower capacity of developing states to take preventative action. Taken together, these limitations mitigate the obligation to one of using all due diligence not to cause *serious* extraterritorial environmental damage (Sands 1995; Birnie and Boyle 1992).

The preventative principle, as manifest in international law, has a notable weakness. As can be seen from the *Trail Smelter* and Stockholm Principle 21 formulations, states are not required to prevent environmental damage to their own territory. Indeed, they are free, subject to any other legal obligations (e.g. the 1992 Convention on Biological Diversity), to allow complete environmental destruction on home soil. The notion that the global environment can be neatly parcelled up into sovereign states, and environmental harm in one state does not indirectly affect the broader environment, is fairly artificial.

It is also fair to say that the preventative principle leaves a number of matters unclear. For example, it is not certain what level of environmental damage shall count as 'serious'. If the criteria for 'serious harm' are agreed upon, evidential difficulties remain concerning causation. In the 1970s the UK notoriously denied the existence of a direct causal link between sulphur and nitrous emissions from its power generation plants and acid rainfall in Scandinavia and Germany (Handl 1990; Yearly 1992). Neither is it unequivocally indicated whether environmental impact assessments are required prior to potentially polluting activities or whether states shall incur liability for failure to prevent environmental harm (Williams 1995, 436).

On the other hand, dilution of the primary obligation to prevent pollution may be no bad thing. Some commentators (e.g. Lis and Chilton 1993; North 1995) consider that society is in danger of too much preventative action. Zero pollution is, they argue, the environmentalists' utopian dream. Zero pollution is not physically possible due to the second law of thermodynamics which requires some waste energy or matter to result from every human activity (Georgesçu-Roegen 1971). What is required according to economists (e.g. Pearce and Turner 1990) is reduction of pollution to, but not beyond, an *optimal* level where the marginal costs of reduction just equal the marginal benefits obtained from such reduction.

EU environmental law and policy embodies the preventative principle in a number of ways. It appears, first of all, at a general level. The EU's Action Programmes on the Environment have stressed the need for preventative action (Jans 1996, 284), and the EC Treaty contains the requirement that environmental policy be based on the principle that preventative action should be taken (Article 174). In *Peralta* [1994] (ECR I-3453) the ECJ was asked to determine, in effect, whether the preventative principle as expressed in Article 174 ought to be interpreted as subject to certain qualifications such as a requirement of efficiency or non-discrimination. The case arose from the prosecution of the master of an Italian vessel for discharging caustic soda outside Italian territorial waters. Italian law made this an offence for vessels registered in Italy but not for those registered in other states. Furthermore, masters of Italian vessels, but not others, were to be penalised by suspension of professional qualifications. The Italian court asked the ECJ whether the principle of prevention in Article 174 precludes Member States from passing laws discriminating against Italian nationals and forcing Italian vessels to use inefficient discharge methods not required by the International Convention for the Prevention of Pollution from Ships. The Court's response was that Article 174 of the EC Treaty is 'confined to defining the general objectives of the Community in the matter of the environment' and that Article 176 guarantees that Member States may introduce measures which are more stringent than those adopted by the EU. Consequently, the preventative principle did not preclude the legislation in question.

The preventative principle is also manifest in more specific EU measures, such as the Directive (EEC/85/337) requiring environmental impact assessment (EIA) for certain projects. This is discussed in greater detail below, in relation to the precautionary principle. Article 2 of Directive 85/337 requires the Member States to adopt

> all measures necessary to ensure that, before consent is given, projects likely to have significant effects on the environment by virtue, *inter alia*, of their nature, size or location are made subject to an assessment with regard to their effects.

The preventative principle exists in one form or another in most nations' legal systems. In common law systems prevention is implicit in the law of 'private nuisance'. The essence of private nuisance is that a person may be prevented from, or made liable for, use of property in such a manner as to cause injury to the property of another. As with the international duty to stop transboundary harm, the duty to prevent harm to neighbours is not absolute, but rather one of 'reasonable use'. As Lord Wright noted in *Sedleigh-Denfield* v. *O'Callaghan* [1940]:

> A balance has to be maintained between the right of the occupier to do what he likes with his own, and the right of his neighbour not to be

> interfered with. It is impossible to give any precise or universal formula, but it may broadly be said that a useful test is perhaps what is reasonable according to the ordinary usages of mankind living in society.

Conversely, an 'unreasonable use' of land will lead to liability even if all proper care is taken to avoid harm (*Cambridge Water Company* v. *Eastern Counties Leather* [1994]). What is reasonable in a town or built-up area may differ from that which is reasonable in a quiet or rural location. Persons must accept

> the consequences of those operations of trade which may be carried on in his immediate locality, which are actually necessary for trade and commerce, and also for the enjoyment of property, and for the benefit of the inhabitants of the town and the public at large.
>
> (*St Helen's Smelting* v. *Tipping* (1865))

The most important preventative aspect of the law of nuisance is the ability of a person whose property is threatened by pollution, or the risk of pollution, to apply for an order – an *injunction* – requiring the polluter to halt or modify the offending activity. An injunction is an *equitable* remedy. As such it is granted only at the discretion of the court. Generally, however, English courts will grant injunctions even where this results in significant costs and inconvenience to the polluter (e.g. *Pride of Derby and Derbyshire Angling Assoc.* v. *British Celanese* [1953] and *Brocket* v. *Luton Corporation* [1948]). Many US State courts take a similar approach (e.g. *Whalen* v. *Union Bag and Paper Co.* (1913)) because the non-availability of an injunction is seen as an infringement of the applicant's property rights. In *Whalen* the court stressed that polluters should realise that the law protects property rights and should avoid discharges that infringe those rights.

One problem with the general rule that injunctions should be available is that they may do more harm (especially economic harm) than good. The polluter may be a major employer or service provider for the area in question. Courts have developed two approaches to deal with this. The first is to award damages *instead of* an injunction (i.e. no preventative effect). In English law (*Shelfer* v. *City of London Electric Lighting Co.* [1895]) compensation will be awarded instead of an injunction if:

1. the injury to the claimant's legal rights is small;
2. it is capable of being estimated in money;
3. it can be adequately compensated by a small money payment; and
4. it would be oppressive in the case to grant an injunction.

A US decision which demonstrates similarities to the *Shelfer* criteria is *Boomer* v. *Atlantic Cement Company* (1970). The defendants operated a large cement factory from which smoke, dirt and vibrations issued forth. Neighbours seeking damages and an injunction sued the defendants. The New York Court of Appeals refused to grant an injunction so long as the defendants paid permanent damages of $185,000 to compensate the claimants for their economic loss. In denying the injunction the court referred to the 'large disparity in economic consequences of the nuisance and the injunction' and observed that the defendant had invested over $45 million in a plant that employed 300 people.

The second technique is to soften the blow of an injunction. This can be done by withholding its actual grant or suspending its operation, giving the polluter time to develop and apply new pollution control technologies (e.g. *Georgia* v. *Tennessee Copper Co.* [1907]). An unusual alternative, available in some US states (but not the UK), is to redress the balance by granting the injunction subject to a requirement that the pollution *victim(s)* pay compensation to the polluter. This technique derives from the case *Spur Industries* v. *Del Webb Development Co.* In the 1950s Webb constructed a retirement community called Sun City. This expanded so that it was close to a cattle feedlot operated by Spur. After a while pollution (flies and odours) from the feedlot caused prospective customers to decline places in Sun City. Del Webb sought an injunction prohibiting Spur's operations. The Court looked at the matter from both sides. On the one hand Spur Industries were causing a nuisance, therefore the court would grant an injunction. But Del Webb were not blameless – by positioning themselves right next to Spur's exiting feedlot they had effectively 'caused' the problem. This was recognised by the court requiring Del Webb to indemnify Spur for a reasonable proportion of the cost of moving their feedlot elsewhere or closing down.

On a strict interpretation, the preventative principle concerns itself with the ability of the law to prevent pollution from arising *in advance*. A special kind of order requiring a landowner to prevent pollution from arising, granted ahead of any actual pollution, is known as a *quia timet* injunction. In English law, to convince the court to grant a *quia timet* injunction, the pollution concerned must be a 'near certainty' (*Redland Bricks Ltd* v. *Morris* [1970]). Possibly, the polluting event must also be imminent (*Lemos* v. *Kennedy Leigh Developments* [1961]; but cf. *Hooper* v. *Rogers* [1975]; see also Silver 1986). Other jurisdictions take a more relaxed view: in some Australian courts the requirement is merely 'there must be a fairly high degree of probability that the nuisance will occur'

(*Kent and Others* v. *James Cavanagh, Minister of State for Works and Others*).

One generic approach to pollution prevention is to focus on best technologies and practices. In the UK, for instance, the Environmental Protection Act 1990 created a system of Integrated Pollution Control which requires the use of the Best Available Techniques Not Entailing Excessive Cost (BATNEEC). BATNEEC implies a stringent approach in which each polluter must achieve the best standards for the industry as a whole.

Another preventative strategy is to grant environmental agencies statutory powers to take action, at the polluter's cost, to prevent pollution from occurring. Section 161 of the UK's Water Resources Act allows the Environment Agency to take preventative steps, and recover costs, if any 'poisonous, noxious or polluting matter or any solid waste matter' is likely to enter any 'controlled waters'.

The main items of preventative legislation in the United States are the Resource Conservation and Recovery Act (RCRA) (Fortuna and Lennett 1987) and the 1990 Pollution Prevention Act. The RCRA, which provides cradle-to-grave regulation of 'hazardous waste', empowers US District Courts to require any party who has contributed to or is contributing to 'an imminent and substantial endangerment to health or the environment' at a waste site to take such steps as are necessary to abate that danger (RCRA section 7003). In 1984 the RCRA was amended to require persons discharging hazardous waste to report the steps that they are planning to take to incorporate pollution prevention methods into their production processes.

The 1990 Pollution Prevention Act came about from Congress's realisation, after twenty years of applying 'end of pipe' solutions to industrial pollution, that it would be cheaper, in many instances, for industrial operators to build pollution reduction measures into the basic design of their operating systems. The Act, which applies to all waste, establishes a hierarchy in which the priority is that 'pollution should be prevented or reduced at the source whenever feasible'. The Act required the administrator of the US Environmental Protection Agency to set up a separate office to serve as administrator under the 1990 Act. Grants can be made, under the Act, to encourage the use of reduction-at-source technologies and techniques. The PPA administrator acts as a clearing house for reduction technology transfers and provides education and support to businesses who wish to adopt a preventative approach (see, e.g., www.epa.gov/opptintr/p2home/p2setup.htm).

Despite the existence of one or two domestic legal examples, most preventative activity is voluntary. Many companies now find that waste minimisation and pollution prevention are aspects of company policy which they wish to pursue, in part to bolster their public image, in part to find cost-effective methods of saving resources and waste disposal costs. The company 3M is a notable example, claiming to have eliminated more than 500,000 tonnes of pollution and saved £530 million in the period from 1975–93 (Lis and Chilton 1993, 51).

The precautionary principle

The precautionary principle is, in essence, the notion that *lack of full scientific certainty* should not prevent or delay action to protect the environment from harm or prospective harm. It is an extension of the preventative principle: prevention should be employed even where the causes and consequences of the environmental peril in question are imperfectly understood.

The origins of the precautionary principle are said to lie in *Vorsorgeprinzip* – a principle of German administrative law (von Molkte 1988; Boehmer-Christiansen 1994) – which translates as 'prior worry or care'. It evolved in political agenda (O'Riordan and Jordan 1995) and through inclusion in numerous international policy documents and conventions. Principle 15 of the 1992 Rio Declaration on Environment and Development gives a typical modern rendition:

> In order to protect the environment, the precautionary approach shall be widely applied by states according to their capabilities. Where there are threats of serious or irreversible damage, lack of full scientific certainty shall not be used as a reason for postponing cost-effective measures to prevent environmental degradation.

Many international environmental laws now encompass a precautionary approach (e.g. the 1982 World Charter for Nature; Ministerial Declarations of the International Conferences on the Protection of the North Sea; and Article 2 of the 1992 Convention for the Protection of the Marine Environment of the North-East Atlantic). The Montreal Protocol to the 1985 Vienna Convention on Protection of the Ozone Layer was amongst the earliest concrete manifestations, setting phase-out obligations for ozone-depleting substances even though, at the time, there was considerable uncertainty about the role of these substances in the ozone-depletion phenomenon.

Ubiquitous reference to the principle has not prevented disagreement concerning its status as a principle of international law. Several commentators (Gundling 1990; Handl 1990; Birnie and Boyle 1992) doubt that the principle has entered customary law, citing lack of consistent state practice and *opinio juris*, and uncertainty as to the concrete substance of the principle. Other commentators (Cameron and Abboucher 1991; Hey 1992; Cameron 1994; McIntyre and Mosedale 1996) disagree, citing the manifold instances in which it is now incorporated into international conventions and judgments.

In EU law the precautionary principle is, as we saw above, one of the bases for community environmental policy (Article 174, EC Treaty). Several EU Directives are premised on the need for action ahead of full scientific certainty. Directive 76/403 requires strict control of disposal of PCBs and PCTs, despite these being only 'widely recognised' to be harmful. Similarly, Directive 78/176 regulates waste from the titanium dioxide industry even though such waste is merely 'liable' to be harmful to human health (Tromans 1995).

EU legislation is one thing. National case law is another (Lavrysen 1998). Some national judges are reluctant to elevate the precautionary principle from one of policy to one of law. Exemplary in this matter is *Leatch* v. *Director-General National Parks and Wildlife Services and Shoalhaven City Council* (1993). Here the applicant appealed against a licence, granted under the National Parks and Wildlife Act 1974, permitting the council to 'take' endangered species as part of a proposed road development. The applicant argued that the Director-General was bound to take the precautionary principle into account in determining the licence application. The judge, Stein J, disagreed. In his view the precautionary principle is merely

> a statement of common sense . . . applied by decision-makers in appropriate circumstances prior to the principle being spelt out . . . directed towards the prevention of serious or irreversible harm to the environment in situations of scientific uncertainty.

Stein J added that although adoption of a cautious approach was 'clearly consistent with the subject matter, scope and purpose of the [1974] Act' it was, nevertheless, not legally obligatory. *Leatch* was followed in *Friends of Hinchinbrook Society Inc* v. *Minister for Environment and Others* (1997), the New Zealand case *McIntyre* v. *Christchurch City Council* (1996), and cited with approval in the UK *Duddridge* case. *McIntyre* turned on the argument that radiation emitted by a proposed radio communications transmitter would be potentially harmful to health and that the New Zealand Resource Management Act contains a precautionary policy. The court accepted much less:

> there *may* be resource consent applications in which a consent authority may consider it relevant and reasonably necessary to have regard to the precautionary principle [emphasis added].

In such cases:

> a consent authority *may* allow its discretionary judgment to grant or refuse consent to be influenced by the precautionary principle to the extent consistent with the statutory purpose of promoting the sustainable management of natural and physical resources and with judicial exercise of that discretion.

A notable exception to this general reluctance is *Vellore Citizens' Welfare Foundation* (1996). This Indian case reasoned that the precautionary principle and the polluter pays principle are 'essential features' of sustainable development which, in turn, is part of customary international law and, thereby, part of Indian domestic law (Anderson 1998). The precautionary principle was also implicit in, although not directly referred to in, the Canadian Supreme Court case *R* v. *Crown Zellerbach* [1988]. This decision determined that the Ocean Dumping Control Act did not require proof of actual harm in its application to cross-provincial boundary waters (Brunnée 1998).

One reason why some national courts have been reluctant to afford independent legal status to the principle is that environmental legislation often already requires a precautionary approach. In the United States the 1970 Clean Air Act, the 1972 amendments to the Federal Water Pollution Control Act, the Federal Food, Drug and Cosmetics Act, the Toxic Substances Control Act, and the Marine Mammal Protection Act all capture elements of a precautionary approach (Bodansky 1994). In cases involving this kind of legislation further reference to an independent precautionary principle could be seen as an unnecessary and complicating factor.

Judges are also sometimes concerned that elevation of the principle to the legal plane may cause evidential difficulties. In *Nicholls* v. *Director National Parks and Wildlife Service* (1994) TALBOT, J observed that

> while [the precautionary principle] may be framed appropriately for the purpose of a political aspiration, its implementation as a legal standard could have the potential to create interminable forensic argument. Taken literally in practice it might prove impossible.

Exploring the precautionary principle

Whether or not courts afford the principle a legal mantle, they still have considerable opportunity to refine its concrete application. Areas which offer the most scope are:

- the range of issues to which the principle should apply;
- the level of risk which should trigger action;
- the type of action required.

The range of issues

Regarding the range of issues caught by the principle, one question is whether it should apply to human health issues, or only to 'pure' environmental risks. In fact, most cases agree that precautionary measures are justifiable not only to environmental problems but also for the protection of human health, even where this infringes free trade rules. In the dispute between the US/Canada and the EU concerning the safety of beef raised using growth hormones (*EC Measures Concerning Meat and Meat Products (Hormones)* the EU sought to justify its import ban on such meat by reference to the precautionary principle. Although the EU ban was ultimately determined to be unlawful under the General Agreement on Tariffs and Trades (GATT) it is notable that the WTO Appeal Panel did not challenge the applicability of the precautionary principle to human health protection. EU cases take a similar line (*Eurostock Meat Marketing Ltd* v. *Department of Agriculture for Northern Ireland*; *R* v. *Medicines Control Agency, ex parte Rhône-Poulenc Rorer Ltd and another* [2000]). In the *Eurostock* case the issue was whether a UK law prohibiting importation of meat products presenting a risk of BSE transmission to Northern Ireland was contrary to EC rules preventing restrictions on free trade. The Advocate-General thought not:

> [a]ccording to . . . the EC Treaty . . . a 'high level of human health protection must be ensured in all Community measures. Consequently, where such measures on the part of the Commission are still pending, it must be open to the Member States to ensure that high level of health protection by taking their own precautionary measures.

Yet the matter is not universally agreed upon. Some domestic interpretations of the principle differentiate between the environment *per se* and human health. In *R* v. *Secretary of State for Trade and Industry, ex parte Duddridge and others* [1994] it appeared to Smith J that:

> [the UK's policy interpretation of the precautionary principle] is intended to protect the environment itself and is not intended to apply to damage to health caused by environmental factors unless those factors are or might in themselves be damaging to the environment in the long term.

The level of risk

Another 'opportunity for clarification' concerns the *level* of risk that ought to precipitate the principle's application. This matter is currently subject to numerous interpretations. Existing formulations of the required risk threshold include (Gullett 1997, 59):

- '*significant* risks' (*ex parte Duddridge*)
- *threats* of serious or irreversible damage (Rio Declaration, Principle 15);
- *reason to assume* that damage is *likely* (1987 London Declaration on the Protection of the North Sea);
- *reasonable grounds* for concern that pollution *may* be caused (1993 Convention for the Protection of the Environment of the North East Atlantic); and
- no proof of *harmlessness* (Decision of the Commission of the Convention for the Prevention of Marine Pollution by Dumping from Ships and Aircraft).

Whichever formulation is utilised, it may be difficult for courts to supervise practical decisions about whether the threshold has been crossed. In some cases evidence of the level of risk will simply be lacking. In *Re Mohr and Great Barrier Reef Marine Park Authority and Marcum and Others*, for instance, neither party had provided evidence of the risk-weighted consequences of a proposed fish farming development. The Authority, therefore, could not be criticised for not taking the precautionary principle into consideration in refusing a permit for the operation.

A legislative response to the evidential deficit is to require risk assessment as part of the process of authorisation. This occurs, for instance, in the EU Directive 90/220/EEC of 23 April 1990 on the Deliberate Release into the Environment of Genetically Modified Organisms. The Directive requires those wishing to release GMOs to present a safety dossier to the EU Commission, including an assessment of the likely risks of release. In *Association Greenpeace France and Others and Ministère de l'Agriculture et de la Pêche and Others* (Case C-6/99) the ECJ determined that the risk assessment arrangements under the Directive were sufficiently stringent to comply with the precautionary principle. This is true even where approval for release is effectively taken by the Commission, and not by the Member State concerned.

In other cases evidence of risk will be present, but inconclusive. As Jans (1996) suggests, decision about the implementation of the precautionary principle may have to be based on 'tentative and indicative scientific data'. To wait for certainty would be contrary to the principle itself. This was the essential point in *Ontario Ltd.* v. *Metropolitan Toronto and Region Conservation Authority* [1996], where the company concerned had been denied permission to infill on lands which lay close to important streams. No model had been presented at the hearing to indicate a threshold for intrusion into the watershed up to which such development could be allowed. In the absence of such a model, the tribunal applied the precautionary principle and determined that development within such land should not proceed. The tribunal's approach was upheld as lawful by the Ontario Divisional Court.

The type of action

A final uncertainty is the type of *action* required on the part of the decision-maker. Does the principle always require abstinence from development or pollution – or does it merely require extra care to be taken in decisions about the permissibility of such activities? In *Greenpeace Australia* v. *Redbank Power Company* (1994) the NGO challenged a consent for a proposed power station, maintaining that scientific uncertainty should not be used as a reason for ignoring the environmental impact of its likely carbon dioxide emissions. Chief Judge Pearlman thought that, even accepting this, the principle does not necessarily entail curtailing the activity in question:

> The application of the precautionary principle [merely] dictates that a cautious approach should be adopted in evaluating the various relevant factors in determining whether or not to grant consent; *it does not require that the greenhouse issue should outweigh all other issues* [emphasis added].

Consequently, the grant of the permit was not overturned.

The competing view is that the precautionary principle requires systematic evaluation and consideration of the risks of action; a process encapsulated by the methodology of Environmental Impact Assessment (EIA). In the EU, the Environmental Impact Assessment Directive (Directive 85/337/EEC as amended by 97/11/EC) requires EIA for projects likely to have significant environmental impacts, as listed in the first two Annexes to the Directive. EIA is compulsory for all Annexe I projects (i.e. major works such as crude oil refineries, power stations, and radioactive waste reprocessing or storage plants). EIA is also required for those Annexe II projects that Member States determine appropriate, taking into account the characteristics, location and potential impact of the projects (as elaborated in Annexe III). The Directive gives Member States discretion to determine which Annexe II projects require EIA. This discretion is, however, not unlimited: in *World Wildlife Fund (WWF) EA and others* v. *Autonome Provinz Bozen* [2000] the ECJ held that national governments must apply the key test of whether the Annexe II projects would be likely to have significant environmental effects because of *their size, location, or nature*. Member States are not permitted to set criteria for assessment based on just one of those (e.g. size alone) – all three must be taken into consideration (*Commission* v. *Ireland* (1996)). Of course, some would argue that determining whether a project is likely to have significant environmental effects is something which cannot itself be determined without an EIA, making the test itself somewhat impotent.

The substantive process of EIA under the Directive consists of three stages. First, the developer must submit an Environmental Statement (ES) outlining the main alternatives studied and the main reasons for the choice of project, taking into account the environmental effects. In the second stage the authority puts the developer's ES out to public consultation. In the third stage the ES and the results of the consultation exercise are taken into account in the final decision to grant or withhold development consent.

In practice the EU's EIA Directive has been a mixed success (Alder 1993). Undeniably, it has been of value in drawing the attention of both developers and development control authorities to the importance of prior research into environmental effects. It does, however, suffer from a number of weaknesses as a manifestation of the precautionary principle.

First, in the early years of its implementation projects escaped EIA on the grounds that they were too small to have a significant environmental impact when, in reality, they were part of a larger scheme that, overall, may have detrimental effects. Fortunately courts have responded to block this loophole. English courts have ruled that development authorities should, in the consent process, ask themselves whether the proposed development is, in reality, an integral part of a more substantial development (*R.* v. *Swale Borough Council, ex parte Royal Society for the Protection of Birds* [1991]). The ECJ has also observed (*Commission* v. *Ireland* (1996)) that

> Not taking account of the cumulative effect of projects means in practice that all projects of a certain type may escape the obligation to carry out an assessment when, taken together, they are likely to have significant effects on the environment within the meaning of Article 2(1) of the Directive [which would be contrary to EU law].

Second, the public consultation requirements may be partly circumvented by developers applying for a determination of whether EIA is required, *prior* to making an application for a development consent. Such pre-application determinations are voluntary and, consequently, do not require public consultation.

Third, the ES is prepared by or on behalf of the developer, not a neutral party, and so is likely to give a distorted picture of the true environmental impact.

Fourth, there is no formal 'scoping' mechanism for determining the range of the considerations which should be included in the EIA process or the range of alternatives that should be considered.

Finally, even a 'negative' EIA does not preclude the grant of a development consent. Assessment is not the same thing as prevention.

EIA in the US is governed by the National Environmental Policy Act of 1970 (NEPA). Section 102(2)(c) NEPA requires, in the case of 'proposals for action and other major Federal actions significantly affecting the quality of the human environment', preparation of a report stating, amongst other things, the 'environmental impact' of the proposal. This requirement has had significant impact on government agencies such as the National Environmental Protection Agency as well as on private firms who require agency licences for their operations. Environmental groups and businesses frequently do legal battle over the vexed question of whether an EIS is needed due to the unfortunate vagueness of terms such as 'significant impact' and 'human environment' (Kubasek and Silverman 1997, 133).

Businesses and the US EPA resist EIA because of the vast amounts of time and money involved. Sympathetic reviews of NEPA conclude that, overall, the Act has been a success, forcing greater environmental awareness and better agency planning (Kubasek and Silverman 2000, 141). A more realistic view, perhaps, is that the Act 'prohibits uninformed – rather than unwise – agency actions' (*Robertson* v. *Methow Valley Citizens Council* (1989)). The courts tend to defer to agency 'expertise' in the conduct and conseqences of concluding an EIS. Meyers (1991) concludes:

> all too often, agencies' actions are not driven principally by scientific expertise. Rather, decisions are made to protect 'vested' economic interests, or because revenues generated by resource exploitation are often returned to the agency, rather than as general funds to the United States Treasury.
>
> (Meyers 1991, 656)

Another uncertainty concerning the principle is whether action is required regardless of cost. It would appear not, since several legal instruments limit the application of the principle to 'cost effective' measures. For instance, the UK's policy (DoE 1990) is that

> the government will be prepared to take precautionary action to limit the use of dangerous materials or the spread of potentially dangerous pollutants, even where scientific knowledge is not conclusive, *if the balance of likely costs and benefits justifies it*. The precautionary principle applies particularly where there are good grounds for judging either that *action taken promptly at comparatively low costs* may avoid more costly damage later, or that irreversible effects may follow if action is delayed [emphasis added].

Overview

Not surprisingly, given its somewhat ambiguous content, the principle has been subject to a number of criticisms:

- it is 'a marvellous piece of rhetoric' which encourages us to organise our lives around predictions that are unlikely to come true (Wildavsky 1995);
- it is too vague to serve as a regulatory standard (Bodansky 1991);
- it does 'little more than put a label on a growing consensus on the approaches to be taken' (Nollkaemper 1991);
- it is ambiguous and, therefore, can result in weak law (Tromans 1995);
- it requires rejection of scientific method in favour of policies based on 'mere suspicion' (Gray 1990);
- it urges inaction, creating more risk than it solves (Cross 1996);
- it is not preferable to a 'trial and error' approach (Brunton 1995);

- it is worse than the legal principle in *Alice in Wonderland* (sentence first, verdict afterwards) since it amounts to 'verdict first, trial afterwards and no need for evidence' (Milne 1993).

Despite these alleged weaknesses, which cannot be evaluated here, it is clear that, overall, the precautionary principle is gradually gathering momentum and moving towards recognition as a legal principle of environmental law. It looks likely to be of increasing importance in this century as a standard against which environmental decision-making is judged.

The polluter pays principle

The polluter pays principle (PPP), which derives from policy of the Organisation for Economic Cooperation and Development (OECD 1975), captures the notion that a polluter should pay for the environmental costs of his or her activities. The original OECD definition was that:

> The principle to be used for allocating costs of pollution prevention and control measures to encourage rational use of scarce environmental resources and to avoid distortions in international trade and investment is the so-called 'Polluter Pays Principle'. The Principle means that the polluter should bear the expenses of carrying out the above mentioned measures decided by public authorities to ensure that the environment is in an acceptable state. In other words, the cost of these measures should be reflected in the cost of goods and services which cause pollution in production and/or consumption. Such measures should not be accompanied by subsidies that would create significant distortions in international trade and investment.

It is now embodied in many instruments of international law. For example, Principle 16 of the 1992 Rio Declaration tells us that

> States should endeavour to promote the internalization of environmental costs and the use of economic instruments, taking into account the approach that the polluter should, in principle, bear the costs of pollution, with due regard to the public interest and without unduly distorting international trade and investment.

Similarly, the Convention for the Protection of the Marine Environment of the North-East Atlantic, Art. 2.2b provides that

> The Contracting Party shall apply . . . (b) the polluter pays principle, by virtue of which the costs of pollution prevention, control and reduction measures are to be borne by the polluter.

There are two clear justifications for this principle: one economic, one ethical (Alder and Wilkinson 1999). The economic justification is explained above. By reflecting the costs of pollution control back onto the polluter ('internalisation') the balance is restored. This balancing effect of environmental liability can act as an adjunct to conventional 'command-and-control' regulatory approaches to environmental harm (Grant 1996) and aims at creation of a uniform and fair world trading system (Birnie and Boyle 1992, 109). Theoretically, the polluter pays principle applies both to (prospective) preventative controls and (retrospective) liability regimes: in both cases internalisation of the costs limits pollution to a socially optimal level (Beckerman 1975). Preventative measures include both public and private costs: polluters, not general taxation, should fund the pollution prevention activities of environmental regulatory agencies.

The ethical argument develops from Aristotle's views on corrective justice: a person who wrongly harms another should give redress for that harm. This protects the established distribution that, in turn, underpins social stability (Heidt 1990, 352).

As with the precautionary principle, the PPP has a somewhat uncertain legal status. It is, as we saw, settled customary law that states should not allow their territory to be used so as to cause damage to the territory of other states. Furthermore, this idea has, since Stockholm Principle 21, extended to unowned areas of the environment. We also saw, however, that the rule is uncertain in content. Consequently, it has never been used as the basis for a major reparation claim between two states for environmental damage. Although there are incidents in which polluters in one state cause damage in another, these matters are usually resolved, if at all, by political diplomacy, by private litigation or by voluntary settlement. For instance, after a fire at a chemical factory in Sandoz, Switzerland, caused the massive pollution of the Rhine in 1986, the company concerned made settlements with those affected amounting to 100 million Swiss francs (Schwabach 1989). A similar instance is the almost complete lack of international law response to the Chernobyl nuclear plant accident.

Specific international conventions have improved on customary law by creating sharply focused PPP regimes for a limited range of substances: nuclear materials, oil, hazardous substances and hazardous waste. The 1969 International Convention on Civil Liability for Oil Pollution Damage places strict (i.e. no fault) liability onto the owners of oil tankers that cause oil pollution damage. Maximum liability rises on a sliding scale according to the tonnage of the vessel unless the shipowner is negligent, in which case liability is unlimited. The International Convention on the Establishment of an International Fund for Compensation for Oil Pollution Damage 1971/1992 meets any portion of oil pollution damage which exceeds the ship-based limit, up to a ceiling of around $80 million (Wilkinson 1993). Liability for damage caused by radioactive installations is governed by the 1960 Paris Convention on Third Party Liability in the Field of Nuclear Energy and the 1963 Vienna Convention on Civil Liability for Nuclear Damage. Both conventions establish strict liability of the operator of a nuclear installation. In the Paris Convention liability beyond a ceiling is met by the state in which the installation is based. More recent liability regimes include the 1996 International Convention on Liability and Compensation for Damage in connection with the carriage of Hazardous and Noxious substances by Sea and the 1999 Liability Protocol to the Basel Convention on the Control of Transboundary Movement of Wastes and their Disposal.

The PPP features in EU law at several levels. To begin with, under Article 174 of the Treaty, EU environmental policy must be based on, amongst other things, the PPP. As we have seen before, these guiding principles of EU environmental law can be utilised as 'benchmarks' for judicial review of secondary legislation. This occurred in *R.* v. *Secretary of State for the Environment and another, ex parte Standley and others* [1999]). Here farmers alleged that the UK's application of the EU Nitrates Directive (Directive 91/676) infringed the requirement that EU environmental law/policy be based on the polluter pays principle. The Nitrates Directive requires restrictions of agricultural practices in regions draining into watercourses suffering from eutrophication (excessive mineral enrichment). The applicants, who owned farms in areas identified by the Secretary of State, alleged that the forthcoming agricultural restrictions would cause them loss of profits and devalue their farms. This, they alleged, infringed the PPP: whilst they alone would bear the costs of the agricultural restrictions, many of the eutrophic sources in question could be attributed to others outside of the designated zones or to non-agricultural origins. The ECJ responded by pointing out that Member States must impose on farmers only the costs for reduction or avoidance of the water pollution caused by nitrates for which they are responsible, to the exclusion of any other cost. So long as that interpretation were strictly applied the Directive would comply with the principle.

As well as existing at the general level, the PPP is referred to in some environmental Directives (Directive 75/442/EEC on waste; Directive

75/439/EEC on the disposal of waste oils; and Directive 1999/31/EC on the landfill of waste). For example, Article 11 of Directive 75/442 requires that

> In accordance with the 'polluter pays' principle, the cost of disposing of waste, less any proceeds derived from treating the waste, shall be borne by: the holder who has waste handled by a waste collector or by an undertaking referred to in Article 8; and/or the previous holders or the producer of the product from which the waste came.

At the pan-European level, environmental liability is facilitated by the Lugano Convention on Civil Liability for Damage Resulting from Activities Dangerous to the Environment. This convention, to which neither the EU nor the UK is yet a party, makes the 'operator' of certain activities – use of genetically modified organisms, and the operation of waste treatment and disposal facilities – strictly liable for any resulting environmental damage. The EU has, for some time, been considering instituting its own system for environmental liability. One means of doing this would be for the EU to become a party to the Lugano Convention. However, a recent EU White Paper on Environmental Liability (COM (2000) 66 final) suggests that a new framework Directive may be introduced, providing strict liability for damage caused by EC- regulated dangerous activities.

Exploring the PPP

Although there are a great many issues of interest surrounding the PPP, two questions stand out as requiring further consideration. These are, first, 'who is the polluter?' and, second, 'what damage should the polluter pay for?'

Who is the polluter?

Definition of 'the polluter' is fairly straightforward in the abstract: 'someone who directly or indirectly damages the environment or who creates conditions leading to such damage' (Council of the European Communities 1975). Concrete applications, however, give a mixed and less conclusive answer. In the United States, CERCLA and the RCRA take very broad approaches to the matter (Roberts 1987). Liability under CERCLA attaches to four classes of potentially responsible parties (PRPs):

Box 4.1

Implementing the PPP in US legislation

The most striking example of 'polluter pays' legislation is that found in the United States. Under the Resource Conservation and Recovery Act (RCRA) the US Environmental Protection Agency can recover the costs of dealing with hazardous waste which is causing, or is likely to cause, environmental damage. This liability affects any party who has contributed to or is contributing to 'an imminent and substantial endangerment to health or the environment'. Liability is strict; that is, there is no need to establish any kind of fault (*United States* v. *Ottati and Goss* (1985)). Liability is retroactive; that is, it applies in respect of damage caused before the enactment of the Comprehensive Environmental Response Compensation and Liability Act (CERCLA), regardless of the fact that the substances in question may have been lawfully deposited at that time (*United States* v. *Shell Oil* (1985)). Liability is 'joint and several'; that is, where there is a range of polluters any one polluter can be fixed with the total costs.

The weakness of the RCRA – that its clean-up provisions apply only to sites which pose an 'imminent hazard' where a financially responsible site owner or operator can be located – led Congress to enact CERCLA. The innovative aspect of CERCLA was the creation of a fund of $1.6 billion – the 'Hazardous Substance Response Trust Fund' – designated for the clean-up of dangerous waste sites. 'Superfund', as it became known, was financed mainly (87.5 per cent) by revenues from the petroleum and chemical industries. Amendments to CERCLA in 1986 increased this fund to $8.5 billion, with a greater share being raised by a broad corporation tax.

The overall effect of CERCLA has not been good. Commentators (e.g. Lyons 1986) have criticised the system on grounds of inequity. Freeman (1986), for example, argues that CERCLA's retroactivity is inconsistent with congressional intent, unconstitutional, and contrary to the clear common law presumption against retroactivity of legislation. Neither has CERCLA been sucessful in clearing up pollution. The cost of remediation per site has been around $25 million. The number of identified polluted sites has continued to grow faster than their clean up under this legislation: in 1998 around 1,500 sites were identified as requiring remedial action, but only 178 sites were totally finished (Kubasek and Silverman 2000, 262). CERCLA has huge 'transaction costs': insurance companies and their policy holders do battle in 'knock on litigation' over who should bear the real costs of cleaning up hazardous waste sites (Gergen 1994, 629).

1 Current owners or operators of the facility, regardless of whether the owner or operator caused the release of the hazardous substance.
2 Past owners or operators, who owned or operated the facility at the time of disposal of the hazardous substance.
3 Any person who arranges for the treatment or disposal of the hazardous substance owned or possessed by that person.
4 Any person who accepts or accepted any hazardous substance for transport to disposal or treatment facilities, incinerators or sites selected by that person from which there is a release or a threatened release.

The courts have cast the net widely in interpreting these criteria. In *New York* v. *Shore Realty* (1985) the current owner of contaminated land was held to be liable where the hazardous substances were deposited on the property before he obtained title. In *United States* v. *Carolawn Chemical Company* (1988) a chemical company that had held title to a site for only one hour was classified as 'owner or operator'. These cases are now mitigated by the availability, through the 1986 amendments to CERCLA, of an 'innocent landowner' defence (Gergen 1994, 649).

In other manifestations of the PPP, the polluter is much more narrowly defined. As we have already seen, in the international oil and nuclear damage conventions, liability is channelled onto one party only. This may be justified since the parties concerned (owners of tankers and nuclear installations) are unlikely to be impecunious and, in any event, are required to carry insurance.

One way of avoiding unjustified breadth in definition of the polluter is to set up a priority system with the most culpable identified first. The contaminated land regime instituted by the UK's Environment Act 1995 illustrates this approach. The 1995 Act allows 'remediation notices', requiring decontamination, to be served on an 'appropriate person'. An appropriate person is, in the first instance, 'any person, or any of the persons, who caused or knowingly permitted the substances, or any of the substances, by reason of which the contaminated land in question is such land to be in, on or under that land'. Only if no such person can be found does the class of appropriate persons open up to include 'the owner or occupier for the time being' of the land in question. Before it was formally abandoned, the EU's proposed Directive on Liability for Damage Caused by Waste (Com (91) 219 Final) had employed the same tactic. Strict liability was to have been placed, initially, on the producer of the waste in question. It would have been open, however, for the producer to identify other parties, who would then have stood in his shoes. These would have been the person who imported the waste into the EU; the person who had actual control of the waste when the incident occurred; and the person responsible for a licensed waste site to which the waste has been lawfully transferred.

The common law allows a variety of individuals to attract liability as a 'polluter'. As mentioned above, the law of nuisance attaches liability not only to those who cause a nuisance, but also, in some circumstances, to those who continue a nuisance begun by a third party. This can amount to a positive duty to take steps to remedy pre-existing nuisances. Negligence is an even more flexible action since it is based around the notion of a 'duty of care' not to cause harm to others. Many people who are loosely connected to a polluting event could find themselves facing liability through a negligence action. In the UK case *Scott-Whitehead* v. *NCB and SWA* the Southern Water Authority found itself liable in negligence for water pollution damage to a farmer's potato crop. It had itself issued the authorisation for the pollution and failed to warn the farmer of its existence.

Oil pollution litigation indicates just how far the negligence action can reach. The *Amoco Cadiz* ran aground off the coast of Brittany on the 16 March 1978. Around 220,000 tons of crude oil owned by Royal Dutch Shell spilled into the sea, forming a slick eighteen miles wide and eighty miles long, causing extensive damage to the delicate ecosystem along the Breton coast. Claims, based on negligence, totalling $2.2 billion were brought in the United States. These included a successful claim against Astilleros, the Spanish designers and builders of the ship (*In The Matter Of: Oil Spill By The Amoco Cadiz Off The Coast Of France On March 16, 1978*). Astilleros had negligently designed the rudder and steering assembly, such that it was prone to failure, hence the ship prone to catastrophe.

What should the polluter pay for?

A perennial question in the implementation of the PPP is the range, type and assessment of costs that the polluter should pay for.

In its early formulations the PPP was considered only to apply to the costs of pollution prevention. This implied recovery of the costs of matters such as administering a pollution control system, issuing licences, monitoring

Box 4.2

Lender liability and the PPP

A particular concern in the implementation of polluter pays regimes is the treatment of institutional lenders. In an effort to hit those with 'deep pockets' the judiciary in the United States have extended the 'current owners' category to include banks and other property financiers (Jarvis and Fordham 1993). This may not be what Congress intended since Section 101(20)(A) of CERCLA provides that a person who holds indicia of ownership primarily to protect his security interest in a site, without participating in its management, is not an 'owner' or 'operator'. However, two landmark decisions of federal District Courts have construed this proviso closely. In *United States* v. *Maryland Bank & Trust* (1986) a mortgagee, having purchased the mortgaged property outright at a foreclosure sale, was held to be acting to protect its investment, rather than its security. The court refused to place the bank in a better position than any other purchaser: to do so would, it reasoned, provide the bank with a windfall by enabling it to purchase a low-price property which might otherwise be unsaleable. In *United States* v. *Fleet Factors* (1990) a lender under a factoring agreement, whose involvement with management did not, in fact, extend as far as day-to-day involvement with the operations of a facility, was determined to be an 'owner'. The critical point here was that the lender could affect hazardous waste management decisions *if it so chose*. The decision in *Maryland Bank* and *Fleet Factors* caused widespread concern amongst American lending institutions, in response to which the EPA promulgated regulations establishing that they will not seek to defray their costs against a lender which 'has the mere capacity, or ability to influence, or the unexercised right to control facility operations'.

Environmental protection may be ill served by exposing lenders to liability for pollution costs. Liability risks may act as a disincentive to the financing of the redevelopment of so-called 'brownfield sites'. The reuse of industrial land is an important element in preventing urban sprawl and limiting impacts on biodiversity. European and UK polluter pays regimes have, therefore, sought, to avoid the US approach. The Explanatory Report to the Lugano convention states that: '[a]n outside person who has made possible or facilitated a dangerous activity, for example by lending funds for investment in the said activity, may not be considered to be the operator, unless he *exercises effective control* over the activity in question'.

The recent EU white paper on Environmental Liability advises that 'lenders not exercising operational control should not be liable'. In the UK's contaminated land legislation a 'mortgagee not in possession' is expressly excluded from the definition of 'owner' (s.78A, Environmental Protection Act 1990). This, however, does not entirely resolve the problem since mortgagees may need to take possession in order to realise their security.

pollution emissions and so on (Smets 1994). Most legal systems provide for this type of charging. In the UK, for instance, the Environment Agency/SEPA are empowered to make schemes for charges in respect of their range of licensing activities and for recovery of certain other costs (s.41 Environment Act 1995). Later formulations of the PPP all assume, in contrast, that environmental damage is to be included. The issue becomes which kinds of environmental damage, and how to value them.

Pollution often causes damage to unowned aspects of the environment. In many cases it is not clear whether these are to be included in statutory or international treaty regimes. After the Chernobyl accident one of the reasons why the USSR, as it then was, refused to pay compensation to affected states was that the relevant definition, in the Paris Convention which governed liability, was 'any loss of, or damage to, property'. The actual damage suffered, in contrast, was essentially pecuniary loss and inconvenience – for example, cattle kept from grazing; milk which had to be transformed into cheese; crops which had to be destroyed (Pelzer 1988).

A particular focus for the debate over the scope of allowable damage has been oil pollution legislation. The Civil Liability and Fund Conventions originally provided that:

> 'Pollution damage' means loss or damage caused outside the ship carrying oil by contamination resulting from the escape or discharge of oil from the ship, wherever such escape or discharge may occur, and includes the costs of preventative measures and further loss or damage caused by preventative measures.

No further guidance was given concerning the words 'loss or damage'. Strictly, this was a matter for national courts to determine. Many national courts were, it appeared, prepared to award damages for injury by oil to general or unowned environmental entities. In the *Patmos* case (collision of Greek tanker *Patmos* and the Spanish tanker *Castillo De Montearagon* in the Straits of Messina) the Italian Court of Appeal emphasised that nothing in the Conventions distinguished between public and private property. Although the oil pollution damage in question was not, 'quantifiable in an arithmetic-accounting sense' it was, nevertheless, 'unquestionably of an economic nature' and had 'economic relevance' to the whole community derived 'from a reduced enjoyment of the property–environment because of its unavailability'.

The oil pollution conventions also illustrate the difficulty of calculating the amount that the polluter should pay. Oil pollution often injures areas that are not of high economic value. How should such pollution be valued? Former Soviet legislation calculated costs according to a somewhat arbitrary formula of a 'sum per cubic metre of water affected' (Wilkinson 1993). The issue of quantification of oil pollution damage costs was crystallised in *Commonwealth of Puerto Rico et al.* v. *The SS Zoe Colocotroni et al.* The tanker SS *Zoe Colocotroni* grounded off the coast of Puerto Rico on 18 March 1973. In order to refloat the ship the captain intentionally discharged 1.5 million gallons of crude oil. This drifted into Sucia Bay, contaminating mangrove swamps, and killing an estimated 92 million small marine animals. The District Court awarded damages of over $14 million for the costs of clean-up, restocking and monitoring. The District Court had based restocking costs on the market cost of replacements, from commercial laboratories, for each and every dead organism. The defendants appealed. The Court of Appeals allowed the appeal, vacating the award on the ground that the plan for rehabilitation was *unreasonable*. It was impractical and excessive to consider purchasing a complete set of replacement organisms. A reasonable plan must give consideration to avoiding duplication of expected natural regeneration, and the extent to which efforts beyond a certain point would become disproportionately expensive. Oil pollution damage quickly self-repairs (Melke 1990). Often nature heals more quickly when left alone than when clean-up and restocking activity is undertaken (Baca *et al.* 1987, 459).

The *Patmos* and *Zoe Colocotroni* cases convinced the oil industry of the need to press for alteration of the convention definitions. In 1992 new Protocols to the Conventions were concluded which now define 'pollution damage' as,

(a) loss or damage caused outside the ship by contamination resulting from the escape or discharge of oil from the ship, wherever such escape or discharge may occur, provided that compensation for impairment of the environment other than loss of profit from such impairment shall be limited to costs of reasonable measures of reinstatement actually undertaken or to be undertaken;
(b) the costs of preventative measures and further loss or damage caused by preventative measures.

This definition makes it clear that the conventions do extend to environmental damage *per se*, but only so far as is 'reasonable', and then only where 'actually undertaken or to be undertaken'. The EU's white paper on Environmental Liability similarly supports the use of a standard of 'reasonable costs of actual restoration' as the primary measure of damages.

spreads the costs over a range of parties, many of whom will never have caused or contributed to any pollution.

- Where a failing company is made liable, it is usually the creditors of the company who stand to lose financially if the company is made to pay for environmental damage. Clearly these persons are unlikely to have had anything to do with the pollution itself. Naturally, the courts are usually unwilling to see such a circumstance (Shelbourne 2000). In *Re Celtic Extraction Ltd (in liquidation)*; *Re Bluestone Chemicals Ltd (in liquidation)* (1994) MORRIT LJ, referring to an EU Directive, observed that 'There is nothing in the directive to suggest that the "polluter pays principle" is to be applied to cases where the polluter cannot pay, so as to require that the unsecured creditors of the polluter should pay to the extent of the assets available for distribution amongst them.'
- The PPP, at least as manifest in CERCLA, amounts to an inefficient system which relies too greatly on the 'deep pockets' of wealthy, potentially responsible, parties (Gergen 1994).
- Polluters who pay usually pass the cost onto consumers. This, however, is not such a disadvantage from a theoretical perspective: the increased cost of the goods concerned sends (or should send) the required incentive signals to the company (Gaines 1991).
- The polluter pays principle can be reinterpreted as 'he who pays may pollute' which, arguably, is an unethical approach to environmental protection.
- There are certain exceptions to the principle for justifiable and limited categories of subsidies (Gaines 1991).
- Even strict application of the PPP may not result in a level economic playing field since control costs vary according to the environmental standards adopted in each state; full economic equity requires uniformity of environmental standards (Gaines 1991).
- As we have seen above, there are many practical difficulties involved in arranging for the internalisation of pollution costs.
- Although economic theory predicts that the principle will contribute to optimal welfare, unless the money raised by the imposition of pollution costs on the polluter is spent on the environment, society will have fewer 'goods' of an environmental nature and more of some other kind.

Despite these observations, the principle is firmly entrenched in environmental policy and law and will act as a formative and guiding principle at every level of legal control in the twenty-first century.

Key points

- The existence of environmental principles at the policy level is not the same as the existence of legal environmental principles. One needs to check carefully, on a case-by-case basis, to see which principles have been imbued with legal character and to what extent.
- The precautionary principle is essentially a response to gaps in human knowledge: it requires action to protect the environment, and possibly other responses, ahead of full scientific certainty.
- The polluter pays principle derives from theories of justice and economics, but predominantly the latter. It requires polluters to cover at least the costs of regulation of their activity and probably also the restoration costs for ensuant environmental damage.

Further reading

O'Riordan, T. and Cameron, J. (1994) *Interpreting the Precautionary Principle*, London: Earthscan.

And for a less supportive viewpoint: Cross, F.B. (1996) 'Paradoxical Perils of the Precautionary Principle', *Washington and Lee Law Review*, vol. 53, no. 3, pp. 851–925.

Discussion questions

1 Why is it important that environmental laws are underpinned by a coherent set of principles?

2 Can the criticisms of the precautionary principle be satisfactorily addressed?

3 Is the polluter pays principle an unnecessary burden on industry?

4 Given nature's capacity to regenerate, why should the PPP require imposition of the costs of restoration of a damaged environment on a polluter?

5 Techniques of environmental law

- 'Direct' or 'command-and-control' regulation
- Self-regulation
- Provision of environmental information and education
- Judicial review and citizen suit
- Environmental protection through property rights

Environmental law is mainly concerned with the promotion of environmentally positive behaviours and the diminution of environmentally harmful behaviours. It cannot achieve these goals simply by existing: it must be *applied* through several legal 'mechanisms' or 'techniques'. The range of techniques that law has at its service includes (but is not limited to) the list above. It is tempting to add 'market instruments' to this list. However, economic approaches – such as the use of taxes, subsidies and tradable permits – are considered sufficiently distinct to deserve independent treatment. It is also unclear that these are legal, as opposed to purely economic, techniques. Economic instruments are, therefore, dealt with in Chapter 6, as part of the general discussion of economics and environmental law.

The main theme of this chapter is that each legal technique has unique strengths and weaknesses. Reliance on a single technique is never wise, and any hope of creating an ecologically stable future will require a prudent mix of each approach. Law can be structured to best facilitate each technique. However, getting the mix right is essentially a political question that governments and political institutions must decide.

'Direct' or 'command-and-control' regulation

The most widespread technique for the application of law to environmental problems is the creation of *standards* for polluting or damaging activities, and the *enforcement* of those standards by environmental regulators. Together, these steps are known as 'direct' or 'command-and-control' regulation.

Standards can be divided into several categories characterised by different forms of legal intervention: emission standards, environmental quality standards, process standards and product standards. Each has certain merits and demerits that will confirm or contra-indicate its suitability for particular environmental problems.

Uniform versus individuated standards

Environmental standards can be uniform, in the sense that they apply consistently across a whole industry or state. They can also be individuated, in the sense that they apply different controls, or differing levels of control, to particular groups or individuals within an industry or state.

Uniform standards have advantages: low costs (derivation, application and implementation); inter-polluter equity; and the facilitation of an economic 'level playing field' (Ogus 1994, 33). The tendency towards uniform standards has been accelerated by the increasingly important role of international standardisation bodies such the International Organisation for Standardisation (ISO) (Roht-Arriaza 1995). This phenomenon is visible in the EU where, increasingly, Directives commonly draw on international standards for the detail of control (Salter 1993). For example, Directive 92/75/EEC on energy labelling draws on the European Committee for Electrotechnical Standardization (CENELEC) standard EN 60705:1995 for methods for measuring the performance of microwave ovens for household and similar purposes. When new laws are introduced industries may proactively press for uniformity in the form of 'general binding rules' in return for less expensive regulatory regimes (ENDS 2000a). Of course, whether such rules are in practice general or binding is another matter (Bell and McGillivray 2000, ch. 7).

Uniform standards, however efficient, are not without their demerits. Standards which are superficially uniform may, in reality, have non-uniform impacts and 'further entrench the dominant sector of large companies by imposing higher relative unit compliance costs on smaller companies' (Yeager 1999, 100). Where standards are set for whole industries, 'regulatory capture' may occur; that is, a regulated industry may be able to exert undue influence over the standard-setting process, resulting in standards that are overly sympathetic to the industry point of view (Ricketts and Peacock 1984).

A related criticism is that uniform standards do not easily take account of differing environmental conditions across states or regions (Faure 1998). Indeed, differing environmental standards (hence differing costs to industry) may be *necessary* if trade is to take place at all. We can add to this the political point that imposed uniformity detracts from local or state sovereignty. In the EU, in an effort to balance harmonisation and Member State sovereignty, the environmental Directives and Regulations are often used merely to create *minimum* environmental standards, which Member States are legally permitted to exceed if they wish (Articles 95 and 175 EC Treaty; Bar and Albin 2000).

Finally, it should be noted that uniform standards are not necessarily *high* standards: prior to the Maastricht Treaty the passage of most EU environmental law required unanimous vote, giving Member States with low environmental aspirations the ability to force dilution of any proposed measure. Voting is now done on a qualified majority basis, presumably lessening this problem.

The main alternative to uniform standards is individuated emission; that is, standards that are able to take into account the financial and technical situations of individual polluters and the impacts of their activities on recipient environments. For instance, the 1994 Sulphur Dioxide Protocol to the 1979 Geneva Long-Range Transboundary Air Pollution Convention takes a *critical loads* approach by setting targets for reduction of sulphur emissions by reference to the *impact* of these emissions on the environments of recipient states. Some states will emit more pollution, or pollution that impacts more seriously on fragile environments, than others. The 'critical loads' protocol requires the UK to reduce its SO_2 emissions by 80 per cent by 2010 from a 1980 baseline: a much higher reduction than that required for many other European states.

The downside of individuated standards is that they are necessarily more costly to set and apply. Where standards are set for individual polluters on a case-by-case basis polluters may plead limitations of time, technology and finances in the hope of softer standards – factors which regulators do take into account (Hawkins 1984).

The UK approach to environmental standard-setting is a pragmatic utilisation of different types of standard for different environmental problems. However, the Royal Commission on Environmental Pollution (Royal Commission on Environmental Pollution 1998; Osborn 1999) has recently argued that it needs to embrace a more *synthetic* approach which 'gives proper weight to scientific evidence, to technical possibilities, to economic analyses, and to people's values, and which brings them all together in a synthesis which can command widespread support and legitimacy' (see also Jewell 1999).

Emission standards

Emissions standards, or 'limit values' as they are otherwise known, are routinely found in all sources of environmental law. In the international context, for example, standards limit the maximum permissible emissions of oil from tankers, exhaust gases from aircraft, and sulphurous gases from large power-generating stations. Emission standards show considerable variability of form and may consist of limits for a whole state, an entire industry, individual polluters, or a combination of all of these.

Emission standards have certain advantages. They are relatively easy to set, monitor and enforce. Polluters will also usually be left with considerable freedom concerning the *means* of meeting the specified limits.

Emission standards have two main disadvantages. First, they will not necessarily achieve any given level of environmental quality (that depends upon such standards being set at an appropriate level and properly enforced). Second, they are less than ideal in situations where pollution derives from non-point (i.e. diffuse) sources, such as nitrates that run off agricultural land.

Environmental quality standards

An alternative to emission standards is the specification of maximum allowable concentrations of pollutants permissible *in the environment itself*. Such specifications, usually known as 'environmental quality standards' (EQS) or 'ambient environmental standards' (AES), are particularly useful in situations in which a significant proportion of pollution derives from non-point sources (e.g. nitrate pollution of watercourses) or a complex array of small polluters that would be complex to regulate individually (e.g. urban air pollution from vehicles). For this reason they are routinely used as the basis for air quality standards; for example, the standards for a range of pollutants including lead, sulphur and particulates set by EC Directive 96/62/EC on Ambient Air Quality Assessment and Management.

Environmental quality standards make the most of variations in the assimilative capacity of environments (i.e. the ability of the environment to 'soak up' and neutralise pollutants). It is this feature which led to the UK's insistence, during negotiation of the old EU Directive on Dangerous Substances in Water (76/464/EEC, now repealed), that compliance with water pollution limits should be possible through either emission standards or an EQS route. The UK based its position on the knowledge that its rainy climate, fast-flowing rivers and long coast line would allow it to discharge more pollutants than its continental European partners and still achieve the same river and estuary water quality. The UK thus complies with the Water Framework Directive 2000/66/EC – the replacement for the Directive 76/464/EEC – by setting ambient water standards (e.g. The Surface Waters (Dangerous Substances) (Classification) Regulations 1992 SI 337). The Environment Agency is placed under a statutory duty to achieve these, as far as it is practicable (Section 84 Water Resources Act 1991).

It is worth noting that the EQS approach does not always favour the polluter: when the recipient environment has a weak assimilative capacity, the use of EQS can be *more* demanding than the use of an emission limit regime.

Environmental quality standards can also be applied in a qualitative way to nature conservation. For instance, parties to the 1940 Western Hemisphere Convention are to keep 'strict wilderness reserves' in a virtually pristine state (Sands 1995, 127). Measures which specify the presence or absence of certain desirable or undesirable flora or fauna (e.g. the EU Habitat Directive) can also be thought of as prescribing certain levels of environmental quality.

Process standards

Process standards, also known as 'specification standards' (Ogus 1994), specify the use of or abstention from certain technologies, materials, or practices. For example, certain international conventions require the use of ‘best available techniques’ or ‘clean production methods’ (e.g.1992 OSPAR Convention, Art. 2(3)(b)). Other laws negatively prohibit certain practices – for example, the use of fishing nets below a given mesh size or the use of driftnets.

National and regional law make frequent use of process controls. In the UK, heavy industry is subject to a regime of Integrated Pollution Prevention and Control (IPPC) requiring the ‘Best Available Techniques’ (BAT) for the prevention or reduction of emissions of listed toxic substances (Bell and McGillivray 2000, 380).

Process standards have drawbacks: they may be overly prescriptive of the manufacturing process itself; they can be difficult to relate to a given environmental outcome; if expressed in terms of particular technologies they inhibit innovation and become quickly outdated. Many process standards – such as BAT mentioned above – avoid this pitfall by expressing standards in terms of a general goal coupled with guidelines for the interpretation of that goal for different classes of operator. In other words they eschew the exact specification of technologies which are to count as ‘best’.

Product standards

Products impact on the environment in three ways: in their production, in their use and in their disposal. In each case, law can either restrict products lacking the desired environmental characteristics or, less intrusively, require labelling to indicate their environmental performance.

Laws dealing with ‘production characteristics’ are useful for dealing with products’ extraterritorial impacts. The US Marine Mammal Protection Act 1972 – a notable example – prohibited the importation of fish or fish products harvested in such a manner as to involve the incidental death or injury of marine mammals (e.g. dolphins) in excess of US standards. Other import bans include prohibitions of timber taken from the unsustainable clearing of rainforests; bans on import of whale and seal body parts; and restrictions on importation of fur products derived from animals caught in leg-hold traps.

Less restrictively, law can assist in the identification of products arising from particularly good or bad environmental practices. The United States’ Dolphin Protection Consumer Information Act was introduced as a counterpart to the Marine Mammal Protection Act to prohibit the application of 'dolphin-safe' labels to tuna harvested by vessels using purse seine nets or drift-netting. Similarly, EC Regulation 2092/91 on the Organic Production of Agricultural Products restricts the use of the label 'organic' to produce grown without the use of artificial pesticides or fertilisers. Exaggerated environmental performance claims for products will, in many countries, breach basic advertising rules (Holder 1991).

EC laws requiring products to have environmentally friendly 'use characteristics' include, for instance, requirements for heaters and boilers to meet minimum energy efficiency levels (Directive 78/170/EEC) and stipulations that drink and food be free from hazardous substances (Directives 80/778/EEC and Directive 76/895/EEC respectively).

Standards requiring products to have certain 'disposal characteristics' are less common, but can be useful in certain cases where the main environmental risk occurs at the end of a product's natural life. Detergents can be required to meet minimum standards of biodegradability (Directives 78/176/EEC, 83/29/EEC and 82/883/EEC). A major recent initiative is the EU End of Life Vehicles Directive 2000/53/EC which requires cars to be constructed so that, upon disposal, they can be almost entirely reused, recycled, or disposed of without hazard to the environment.

Despite their obvious utility, product standards have an uneasy relationship with technology. Law, in most cases, follows rather than leads technology. Pollution standards are set by reference to that which is technically achievable and only tightened incrementally as industry itself develops better abatement devices and techniques. This is exemplified by the gradual tightening of standards for operational oil pollution from tankers which was only made possible as the oil companies themselves developed new apparatus (e.g. improved water/oil filters) and strategies (such as the 'load on top' technique) (M'Gonigle and Zacher 1979). Governments, generally, are wary of prescribing mandatory technologies since this locks industry into certain suppliers or, in the case of processes, owners of intellectual property rights, and thereby diminishes incentives to innovate.

Occasionally the more radical option of 'technology forcing' laws may be necessary; that is, laws which require attainment of product or process standards that are not currently feasible. This approach may be thought appropriate if there is evidence that industries are collaborating in 'trusts' to avoid the launch of new technological innovations. For instance, in the late 1960s the US Justice Department brought anti-trust actions against the big-six American car manufacturers alleging a conspiracy to suppress the development of clean-engine technologies (Yuhuke 1994). This ultimately led to Section 202(b) of the US Clean Air Act (1970), which required vehicle manufacturers to achieve reductions of 90 per cent of hydrocarbons and carbon monoxide from light vehicles by 1975.

The success or failure of technology forcing laws can depend greatly on the attitude taken by the courts to their interpretation. Again, the US experience is valuable here. At the time of their inception there was some fairly broad support for the view that the law requiring a 90 per cent reduction in vehicle emissions was speculative and, if rigorously enforced, could cause the collapse of the whole US automotive industry. Consequently, the courts sympathised with the car manufacturers and took a narrow, if not hostile, approach to their interpretation (e.g. *International Harvester Co. Ruckelshaus* (1973)). In the case of SO_2 reductions for power stations, also required by the 1970 Clean Air Act, the Supreme Court was less sympathetic: in *Union Electric Co.* v. *EPA* the power generator's argument that the required standards were economically and technologically infeasible was brushed aside by Rehnquist J who concluded:

> Technology forcing is a concept somewhat new to our national experience and it necessarily entails certain risks. But Congress considered those risks in passing the 1970 Amendments and decided that the dangers posed by uncontrolled air pollution made them worth taking. Petitioner's theory would render that considered legislative judgment a nullity, and that is a result we refuse to reach.

There is also a tension between product standards and free trade laws. Any restriction on the production, use or disposal of a product creates, directly or indirectly, effects on the freedom of manufacturers to import. Thus, unless very carefully constructed, product or process controls can fall foul of laws designed to protect freedom of inter-nation trade such as the GATT agreement – for example, bans on furs from inhumanely trapped animals (Feddersen 1999) and bans on beef reared using artificial growth hormone (Douma 1999). Even product information laws, such as the EU's Eco-labelling Directive, run the risk of incompatibility with international rules on free trade, especially the World Trade Organisation's Agreement on Technical Barriers to Trade (Driessen 1999; Herrup 1999).

Standard setting

Standards are the mainstay of environmental law. It is important, therefore, that they are set in an appropriate manner and with sufficient stringency to perform the task in hand.

A common approach is the use of *thresholds* (Winter 1996a). Ideally, science indicates the threshold for a pollutant at which disease or environmental damage begins to occur. The legal standard is then set at that level or as close to that level as is achievable. In the case of pollutants which are known to have no safe level (e.g. radioactivity), or for which the safe level is unknown, the usual approach is 'As Low as Reasonably Achievable' (ALARA).

Unfortunately the 'ideal' model does not often apply. First, science may not produce reliable objective indications of safe thresholds. Science is not a value-free, neutral, objective methodology (Pepper 1984, 60) and, in many cases, can only give indications of probability rather than absolute guarantees. The use of studies of questionable methodological reliability; difficulties in understanding the connections between cause and effect; the variable and value-laden use of statistics; and questionable assumptions about exposure, inhalation or ingestion, all combine to frustrate efforts to find reliable safety levels (Winter 1996a).

Respectable scientific conclusions rarely pass directly into legal standards but must first pass through the filters of economic and political evaluation. Economic evaluation often struggles with the question of costs: how to compare relatively certain scientific information about effects and costs of abatement with relatively uncertain valuations of the activity in question or its environmental impacts. This cost–benefit analysis is compounded by the fact that, for many pollutants, abatement costs rise exponentially: initial reductions can be achieved easily, at low cost, whereas elimination of the last few per cent units can normally only be obtained, if at all, with difficulty and at great cost. Thus, the 'abatement cost per life saved' rises sharply. The question of how much to value each life saved is clearly political or ethical in nature and is not blessed with an easy solution. Valuing environmental effects is even harder, given the wide range of potential methods for economic valuation of environmental damage (Winpenny 1991).

The actual process of standard setting in the UK differs between uniform standards set for an industry or section of society, on the one hand, and individuated standards set on a case-by case basis for individual polluters, on the other. In the case of uniform standards, primary environmental legislation empowers the Secretary of State to issue Regulations, normally in the form of a Statutory Instrument, setting detailed standards in relation to some aspect of the environment. It is assumed, or may be explicitly required, that these powers will be utilised only after taking into account the views of the relevant statutory bodies, industry and the general public. However, it is uncommon for the primary legislation to specify procedures. As a result it is difficult for NGOs to challenge standards in court by way of an action for judicial review.

Standards set in this manner tend to have a relatively low political profile. They are seen as politically neutral: informed mainly by science and engineering data with a tinge of public opinion. By utilising what Farber refers to as 'striking while the iron is cold' (1992, 68) (i.e. acting after the initially high levels of public concern have died down) agencies can dilute the rhetoric of the enabling legislation and introduce measures that are less burdensome to economic interests, hence easier to enforce, than might be expected. There are other winners and losers from the technical approach to standard setting. The system is biased in favour of those parties 'possessing the financial and technical resources necessary to negotiate with agency decision-makers' (Yeager 1999), but the public may be disenfranchised.

The other type of standard employed in the UK is the case-by-case approach, allowing the Environment Agency or other bodies (e.g. local authorities) to set emission levels for particular polluters at individual sites, incorporated into a site licence. The Environmental Agency is now moving away from this towards a more uniform 'template' approach due to the costs involved (ENDS 1999c). Individual permits remain, however, amongst the most common of environmental techniques.

The standard-setting process in the United States is more open, structured and litigious than that existing in the UK. US environmental standards tend to be generated through the application of formalised risk-assessment and cost–benefit methodologies. These are embedded in 'informal', 'formal' or 'hybrid' rules generated by the US Environmental Protection Agency (Kubasek and Silverman 2000). Informal rule-making requires publication of the proposed rule in the *Federal Register*, followed by a period for submission of written comments by interested parties. After consideration of the comments, the EPA publishes the final rule plus a statement of its basis and support, again in the *Federal Register*. Formal rule-making follows the same pattern but also requires the agency to hold a public hearing at which witnesses argue the merits and demerits of the rule, and are extensively cross-examined on their evidence. Hybrid rule-making attempts to combine the best features from both systems: submission of written comments followed by a less formal public hearing with more restricted cross-examination opportunities. In all three classes, following its final publication in the *Federal Register*, a rule becomes law.

In practice the US rule-making process is beset by problems of legal challenge and delay. Around 80 per cent of agency rules are challenged by judicial review. In an attempt to overcome this problem the EPA has adopted *negotiated rule-making* (Dalton 1993). This involves, first, conduct of a 'convening study' to assess the chances of successful negotiation given the issues involved; second, selection of a balanced and representative negotiating committee; third, engaging the services of a professional mediator; fourth, conducting the negotiations to a successful outcome. Very few negotiated rules are subject to judicial challenge. Delays in publication of standards occur partly because of the defensive approach that the EPA must take given the likelihood of judicial challenge, because of internal disagreements concerning the type and level of standards which should be set and because of lack of agency resources and competence (Farber 1999, 301).

Enforcement of standards

Standard setting is only the first stage of environmental control: standards have to be implemented and enforced. Enforcement styles vary between legal systems (Vogel 1986; Richardson 1982). In fact the British regulatory style, by comparison to the United States, is conciliatory, pragmatic and heavily reliant on discretion (Vogel 1986; Jasanoff 1991). In England and Wales the Environment Agency prosecutes only 1–2 per cent of offences. Hawkins (1984) explains that although regulatory agencies have a visible mandate to secure protection of the environment, they also have the 'invisible' task of balancing environmental and economic values. They cannot, consequently, afford to be portrayed as the 'enemies of business' through strict sanctioning of every small legal transgression. Furthermore, standards are set which are ostensibly strict, in the full knowledge, on both sides, that they will not be rigorously enforced. As Hawkins (1984, 27) puts it '*non-compliance with standards is thus organizationally sanctioned*'.

In the UK resort to prosecution has, historically, been taken as a sign of personal and institutional failure: the field man, or his region of the agency, lacks the necessary expertise and strength to bring about a desired reduction in pollution without resort to the external assistance. Agencies wish to retain their aura of invincibility; thus breaches which are less than certain to result in conviction are rarely brought to court. Agencies, having a long-term relationship with their regulatees, have a vested interest in not escalating a problem that can be resolved through cooperation into an adversarial situation. Prosecution for 'accidental' pollution in serious pollution incidents can prejudice the future willingness of polluters to disclose information about the causes of the accident (ENDS 1999b). Taken together, these factors result in British environmental regulators restricting prosecution to cases involving very serious environmental harm, or cases in which the polluter has been reckless, persistent, malicious or obstructive (Environment Agency 1998).

Enforcement in the United States is more deterrent-based and, consequently, less conciliatory (Bardach and Kagan 1982). Creation of a separate Environmental Crimes Unit within the US Department of Justice and an Office for Criminal Investigations in the EPA, along with improved funding, have, in combination, increased the use of criminal sanctions (Kubasek and Silverman 2000, 20). Despite these factors there is still a serious gap between the standards as published and the situation on the ground. Many environmental laws fail to be properly or fully enforced. US environmental statutes often give state authorities the key enforcement role, with the EPA having a merely supervisory remit. States often fail to enforce rigorously in the knowledge that the EPA is unlikely to sanction the lack of state action.

Enforcement strategies may differentiate between classes of polluter. There is some qualitative evidence to suggest that in the UK environmental regulators take a more lenient approach to small or impecunious polluters (Hawkins 1984). On the other hand, an empirical study found no correlation between the strength of a firm and the likelihood of an enforcement action by United States' EPA (Yeager 1999). Differences in regulatory style can also appear within states, between regulators of different industries and processes, between regions of a state (Shover *et al.* 1984), or even (as in Dutch provinces) within different branches of the same office (Aalders 1993).

In recent times enforcement has become a more transparent matter through the publication of official enforcement policy documents. The Environment Agency has published its prosecution policy (1998) which states that, in addition to the need for sufficiency of evidence, it will consider a number of public interest factors in deciding whether or not to prosecute, including the environmental effects of the offence, its foreseeability, the intent/attitude/history of the offender, and the deterrence effect of prosecution.

Box 5.1

Self-regulation in North America

North American self-regulation consists principally of the US EPA's 33/50 and XL programmes and the Canadian ARET scheme. Under the 33/50 programme a wide range of manufacturing firms voluntarily agreed to reduce their emissions of seventeen large-volume hazardous chemicals by 33 per cent by 1992 and by 50 per cent by 1995 (ENDS 1993). The EPA issued formal 'Certificates of Appreciation' which the participating company could use to enhance its corporate image, promote products and gain a competitive edge in contract bidding. The XL programme consists of agreements with individual companies to decrease their emissions below the levels that are statutorily required, in return for flexibility on the part of the EPA in allowing cross-media and cross-pollutant trades and relaxed enforcement of minor violations of environmental standards (Ginsberg and Cummis 1996; Steinzor 1998). This leads one somewhat cynical commentator to coin the motto 'If it isn't illegal, it isn't XL' (Steinzor 1996).

In the Accelerated Reduction/Elimination of Toxics (ARET) scheme, the Canadian government challenged industry to take voluntary action to achieve, by the year 2000, a 90 per cent reduction in releases of persistent bio-accumulative toxics, as well as a 50 per cent reduction in releases of 117 substances. Both programmes have apparently been successful (Munn Internet ref).

Integrated Pollution Control (IPC) for heavy industry under the Environmental Protection Act 1990 was accompanied by an informal system of self-monitoring in which the Environment Agency decreased the number of site visits and began to rely on companies themselves for monitoring and providing performance data (Reeds 1994). This trend is now likely to be extended to other areas such as water pollution (ENDS 1999c).

One potential objection to self-monitoring is that information generated may be used against the company concerned in criminal action, potentially contravening the 'privilege against self-incrimination' which in many states is a constitutional guarantee. In the US corporations cannot object to this use of the information, since the privilege is limited in scope to personal applications (*Hale* v. *Henkel* (1906)). In the UK neither individuals nor corporations can object, although this is due to a different line of reasoning. In *R.* v. *Hertfordshire CC Ex p. Green Environmental*

Industries Ltd [2000] the local waste regulation authority found and disposed of a large quantity of waste on Green's land. Subsequently, they served Green with a request for information about the waste under the Environmental Protection Act 1990 s. 71(2)). Green Environmental Industries, who were not licensed to keep waste on the site, refused to reply without confirmation from the council that their replies would not be used against them in a prosecution. Their argument was that without this confirmation the request amounted to deprivation of the right to silence: a right enshrined in Article 6(1) of the European Convention on Human Rights which affords a privilege against self-incrimination. The House of Lords did not agree: they dismissed Green's appeal, reasoning that Article 6(1) is concerned only with the fairness of a trial and not with pre-judicial inquiries. The authority was thus quite entitled to request factual information, particularly in view of the urgent need to protect public health from an environmental hazard, even if potentially incriminating.

Self-monitoring is increasingly done through the use of automatic and remote telemetry equipment. In February 1993 the UK's National Rivers Authority (now the Environment Agency) obtained its first prosecution based on data gathered by a remote sampling device, 'Cyclops' (Howarth 1997). The courts will generally accept evidence consisting of automatically gathered data, although it remains open for a defendant to prove that the machine was not working properly at the relevant time.

A logical extension of the above trend in remote monitoring is to rely on evidence of environmental conditions gathered by satellite monitoring. Although this type of evidence must be carefully checked for reliability, there is no reason why, in principle, satellite remote sensing data should not be used to check compliance with environmental law. The US EPA has used satellite data, confirming extensive wetland destruction, as the basis of an out-of-court settlement for the restoration of 8,500 acres of the wetland in question (Hatch Hodge 1997, 703). The EU has also introduced satellite monitoring of fishing vessels as part of its reform of the EU Common Fisheries Policy, to enable location of vessels fishing illegally and to allow verification of the time at sea of licensed vessels. In the future increases in the resolution of satellite imagery and decreases in the expense of monitoring technology may elevate the role of this technique, although there remains the possibility that privacy laws may establish limits to involuntary monitoring (e.g., in the U.S., *Katz* v. *United States*).

through environmental audit will not be subject to prosecution. The EPA's policy, while broadly welcomed by environmentalists, is less than thoroughly endorsed by corporate interests, who are concerned that the policy does not provide *absolute* immunity from prosecution (e.g. an action by a third party) (Kubasek and Silverman 2000, 22–4).

A further response to the courts' reluctance to create an evidentiary privilege has been enactment, by a majority of US states, of privilege legislation (see, e.g. Mazza 1996). A similar trend is emerging in Australian environmental legislation (Bowman 1997). The UK position for information revealed by audits is unclear.

Provision of environmental information and education

Ultimately, the creation of an environmentally sustainable society requires changes in social and personal values and fundamental perceptions of nature. Law can facilitate this process directly, by facilitating the acquisition of environmental information and by promoting environmental education. Increasing public access to environmental information enables people to play a more effective role in decentralised (i.e. local) environmental management which is widely recognised as a prerequisite for sustainable development (e.g. Brundtland 1987, 65).

International institutions involved in environmental information provision include the United Nations Environment Programme's Global Environmental Monitoring System (GEMS) and the International Referral System for Sources of Environmental Information (INFOTERRA). GEMS acquires and analyses environmental information and disseminates the results to national governments. INFOTERRA is a decentralised network of national environmental information sources covering over 8,400 registered sources in seventy-six countries (Nanda 1995b, 639). UNEP also maintain the International Register of Potentially Toxic Chemicals (IRPTC) which gathers, processes and issues information on the properties of hazardous chemicals.

Most developed states or regions provide legal rights of access to environmental information. Such laws enhance the possibilities for the public to become directly involved in environmental protection, provide data that can form the basis for legal challenge of administrative decisions, and allow monitoring of progress (or the lack of progress) in environmental conditions.

The gathering of environmental information at the EU level is largely the responsibility of the European Environment Agency (http://org.eea.eu.int/). In the EU provision of environmental information for the public currently takes place under EC Directive 90/313 on Freedom of Access to Information on the Environment. This requires public bodies to make environmental information in their possession available, on request, to any person, without the need to show any interest or other basis. Public bodies can make a reasonable charge for the information and have up to two months in which to provide it.

Although it is a big improvement on the previous patchy and inadequate network of European information provision laws, the Directive is acknowledged to have not worked well in practice (ENDS 1996; Bell 1997, 174). There are a number of inadequacies. Information held by private organisations is not included. A range of exclusions and discretionary exclusions place much environmental information off-limits, the most important of which apply to (a) information relating to international relations, defence or public security and (b) information affecting commercial confidentiality. There is no provision for information about the information: thus making it difficult for interested individuals to know what to ask for and to whom to address their requests. There is also a question of what Winter (1996b) terms 'operative information' – that is, information about the background to and origins of the environmental information sought. Operative information can be vital for legal actions and generally enriches the understanding of the recipient. In systems such as the USA, Canada and Sweden, where the whole file is handed over, this becomes transparent. The EU Directive, however, does not require provision of operative information and many Member States (e.g. the UK) do not require its provision.

Because of these criticisms, and the desire by the EU to ratify the 1998 United Nations Aarhus convention on public access to information, participation and justice in environmental affairs, the EU Commission has announced plans to strengthen the Directive. This will involve increasing the number of public bodies required to provide information, and narrowing the scope for refusing requests. The new disclosure rules will apply to *all* public authorities that affect the environment, such as transport and energy agencies (currently exempt in some EU Member States) as well as private utilities, such as energy companies, if they 'carry out similar services to public authorities'. Under the new Directive, refusal to provide information will require authorities to show that disclosure of environmental information would have 'adverse effect'.

Authorities will have to weigh the public benefits of disclosure and non-disclosure. The new Directive will also require environmental information provision via the Internet, and through regular 'state of the environment' reports.

Other Directives contributing to the corpus of environmental information at Community level include: the Chemical Substances Directive (Directive 92/32/EEC – declaration of process and product-related dangers); the 'Seveso' Directive (Directive 82/501/EEC – provision of information to the public about risks arising from chemical plants); the Environmental Impact Assessment Directive, and the Eco-labelling Regulation 880/92/EEC.

The United States is blessed with an extensive right to environmental information through the 1986 Emergency Planning and Community Right to Know Act. The Right to Know Act aims to encourage and support emergency planning efforts for hazardous chemicals and to provide citizens and local government with information about potential local chemical hazards. To these ends, plant operators are required to submit to state and local authorities material safety data sheets, hazardous chemical lists, and data on toxic chemical usage, manufacture, and release (Blomquist 1990a). Public Access to this information is assisted by citizen rights of participation in formulation of emergency response plans, and by easy public access to local facility records and the EPA's computerised Toxic Releases Inventory (TRI). The TRI requires firms in certain classes (SIC codes 20 to 39 inclusive) with ten or more employees to submit emissions data for listed chemicals. This data is then open to public inspection (EPA: Internet ref). Empirical evidence shows that adverse TRI emissions information can lead to stock market devaluation which, in turn, puts companies under pressure to improve environmental performance (Konar and Cohen 1997).

The UK's Environment Agency operates a similar, but more limited, 'Chemical Release Inventory' (CRI) which publically lists emissions from major polluters. Criticisms of inconsistency, poor scope and dissemination and late publication of figures (ENDS 1999f) have led the NGO Friends of the Earth to set up its own alternative website (FoE 2000). Apart from the CRI and EU Directive 90/313, environmental information in the UK is available through the numerous public registers of applications, licences and monitoring results which are created and maintained under most of the environmental statutes (Bell 1997, 162). Unfortunately, empirical evidence suggests that the public make little use of these registers (Bell 1997, 175). Whether this is due to apathy or ignorance of the existence of the registers is not clear.

Generally, UK 'access to information' laws are not, as yet, well developed. Historically, UK governments have operated within a 'secrecy culture', buttressed by the Official Secrets Act 1911–39 (Turpin 1985, 464). The judiciary have thought it important to protect the workings of government from public inspection (Lord Wilberforce in *Burmah Oil Co. Ltd* v. *Bank of England* [1980]) and taken the view that the public interest in access to information must be subjugated to the need to preserve confidentiality (*British Steel Corporation* v. *Granada Television Ltd.* [1981]). Some legislative progress was made with the Public Records Act 1958 and the Data Protection Act (access rights for computer-held information). Further non-statutory progress has been achieved under the Code of Practice on Access to Government Information (http://www.homeoffice.gov.uk/foi/ogcode981.htm). The Code generates a presumption in favour of release of information, except where disclosure would not be in the public interest.

The old attitudes to information will, it is hoped, change now that the UK is in the process of passing a Freedom of Information bill. This bill will replace the Code of Practice, and create broader statutory rights of access covering a wider range of public authorities, including local government, National Health Service bodies, schools and colleges, and the police. Access will be regulated by a Commissioner, to whom the public will have direct access. The bill will enable access to documents, or copies of documents, as well as to the information itself.

Environmental education is a more proactive strategy than creating rights of access to environmental information. One should not forget that environmental treaties and legislation themselves form part of the background cultural factors that have a pervasive educational effect (Thorton and Beckwith 1997, 20). Requirements to provide environmental education are increasingly common in environmental treaties and conventions. For example, Article 13 of the 1992 Convention on Biodiversity requires that contracting parties shall:

(a) Promote and encourage understanding of the importance of, and the measures required for, the conservation of biological diversity, as well as its propagation through media, and the inclusion of these topics in educational programmes; and
(b) Cooperate, as appropriate, with other States and international organizations in developing educational and public awareness

programmes, with respect to conservation and sustainable use of biological diversity.

Similarly, Article 6 of the UN Climate Change Convention 1992 requires parties to

(a) Promote and facilitate at the national and, as appropriate, subregional and regional levels, and in accordance with national laws and regulations, and within their respective capacities:
   (i) the development and implementation of educational and public awareness programmes on climate change and its effects;
   (ii) public access to information on climate change and its effects;
   (iii) public participation in addressing climate change and its effects and developing adequate responses; and
   (iv) training of scientific, technical and managerial personnel.
(b) Cooperate in and promote, at the international level, and, where appropriate, using existing bodies:
   (i) the development and exchange of educational and public awareness material on climate change and its effects; and
   (ii) the development and implementation of education and training programmes, including the strengthening of national institutions and the exchange or secondment of personnel to train experts in this field, in particular for developing countries.

Bodies such as the United Nations Institute for Training and Research assist developing countries in training and education as part of their general assistance remit (Phipps and Beck 1997). Some environmental NGOs take part in educational initiatives which include the dissemination of information about environmental law (Casey-Lefkowitz 1993).

Many states now legally require minimum standards of environmental education. In the USA the National Environmental Education Act of 1990 requires the EPA to make environmental education a priority through various activities administered by its Environmental Education Division. The Act supports various environmental education initiatives: an 'Environmental Education and Training Programme'; the award of grants to support environmental education projects, internships for college students and fellowships for college teachers; and a requirement to establish a National Environmental Education Advisory Council (http://www.agiweb.org/legis105/neea.html).

The UK, as is so often the case, lags behind this state of development. In 1988 environmental education was designated as a cross-curricular theme for primary and secondary education in England and Wales. In 1993 a committee appointed by the Department for Education and the Welsh Office produced *Environmental Responsibility: An Agenda for Further and Higher Education*, designed to promote environmental education in the further and higher education sector. However, as Palmer comments (1999), 'statutory timetabled content of school curricula may have squeezed [environmental education] out, whilst the recommendations of Environmental Responsibility for further and higher education remain unfunded and therefore largely unimplemented'.

An important aspect of environmental education is education about environmental law itself. There are, of course, numerous books, courses and journals to which interested parties can refer to increase their knowledge in this matter. Apathy, lack of time or resources or other factors can prevent this from taking place so it is important that governments take all practicable steps to publicise the existence of their own legislative output. Much more can be done in this respect in most countries, especially with the use of the Internet and public libraries. Even where the public are aware of the existence of laws concerning the environment, and have access to them, problems of understanding may remain. The UK's response to this is to issue 'explanatory notes' which explain the form and content of the legislation to readers who are not legally qualified and who have no specialised knowledge of the matters dealt with.

In certain developing states the judiciary have acted in a quasi-political manner to generate pressure for environmental learning. In *M.C. Mehta* v. *Union of India & Others* [1988], for instance, the Indian Supreme Court took the view that, in order to promote the recovery of the River Ganges,

> it is the duty of the Central Government to direct all the educational institutions throughout India to teach at least for one hour in a week lessons relating to the protection and the improvement of the natural environment including forests, lakes, rivers and wild life in the first ten classes. The Central Government shall get textbooks written for the said purpose and distribute them to the educational institutions free of cost. Children should be taught about the need for maintaining cleanliness commencing with the cleanliness of the house both inside and outside, and of the streets in which they live. Clean surroundings lead to healthy body and healthy mind. Training of teachers who teach this subject by the introduction of short term courses for such training shall also be considered. This should be done throughout India.

Similarly in *M.C. Mehta* v. *Union of India* [1992] the court ruled that licences for all cinema halls, touring cinemas and video parlours should include a condition that they should 'exhibit free of cost at least two slides/messages on environment in each show'. I don't know, however, whether this stipulation has been implemented or how effective it has been.

Judicial review and citizen suit

An important technique for environmentalists is the possibility of asking a court, within its supervisory jurisdiction, to check that a decision made by a public body has been made in the correct manner. This is known as the action for 'judicial review'. The action is subject to four important limitations.

First, it is only available in respect of the decisions of *public* bodies. It can sometimes be difficult to know whether a body is 'public' or 'private'. In *R.* v. *National Trust, ex parte Scott and Others* [1997] an application for judicial review was made in relation to a decision by the National Trust to impose a ban on hunting with hounds on its land. The application was turned down: the National Trust is both a charity and a statutory body created and sustained by the National Trust Act 1907. In adopting the ban the Trust was acting as a charity, not as a public body.

The second limitation is implicit in the very nature of the action for review. The courts have repeatedly emphasised that their role is to supervise *procedural propriety* and not the rights and wrongs of the actual decisions made. In other words, the courts will not ask themselves whether the decision is a correct one, but only whether it was correctly made. In order to be correctly made a decision must not be ultra vires (i.e. outside the powers granted to the agency, implicitly or explicitly), by Parliament. It must also be made *reasonably* (*Associated Picture Houses Ltd.* v. *Wednesbury Corporation* [1948]); that is, by avoiding decisions that no sensible person could have reached (*Council of Civil Service Unions* v. *Minister for the Civil Service* [1985]). A reasonable decision takes account of all relevant considerations and is based on no irrelevant considerations.

The third limitation is the fact that a person cannot challenge a decision of a public body in a national court unless he or she has a right of hearing, referred to in legal parlance as a right of 'standing' or *locus standi*. In this matter a balance must be maintained. On the one hand, courts tend to believe that it is their duty to block 'mere busybodies' who would otherwise clog up the legal system. On the other hand there is value in the quasi-regulatory role of interested and well-informed citizens. Indeed, Principle 10 of the Rio Declaration reminds us that

> Environmental issues are best handled with the participation of all concerned citizens, at the relevant level . . . Effective access to judicial and administrative proceeding, including redress and remedy, shall be provided.

United States' courts have moved from a liberal to a strict approach to standing. The 'high ground' was a result of the case *Sierra Club* v. *Morton* (1972), in which the plaintiff challenged a decision of the Forest Service approving a plan for a $35 million resort in the Mineral King Valley – an area of outstanding natural beauty (Stone 1972). The Supreme Court held that the NGO's 'special interest' in the area affected was not enough to ground a right of standing: the NGO must demonstrate 'injury in fact'. Although the challenge in *Morton* was unsuccessful the case set a precedent which, paradoxically, was advantageous to environmental groups since it turned out to be fairly easy for NGOs to show 'injury in fact' by demonstrating loss of recreational amenity (Findley and Farber 1992, 3). In *United States* v. *Students Challenging Agency Procedures* (1973) the Supreme Court helped matters further by adding that standing is 'not to be denied merely because many people suffer the same injury' and that 'injury in fact' can be small so long as it is qualitatively appreciable. The high point of the rules on standing in the US was probably *Duke Power* v. *Carolina Environmental Study Group* (1978) which held that the mere likelihood of the future application of a possibly unconstitutional Act to the plaintiff was a sufficient basis for standing.

Things went downhill in *Lujan* v. *National Wildlife Federation* (1990) (see Note 1991). The NGO in question wished to challenge a decision of the Bureau of Land Management (BLM) reviewing an earlier decision to withdraw federal lands from resource development. Specifically, the NGO wished to allege that the BLM had violated statutory requirements in several respects, and that BLM's actions would open up the land to mining activities. The court denied the NGO standing on two grounds. The lands used by the members of the NGO, on which they based their claim of 'injury in fact', were not actually federal lands, but only lands 'in the vicinity' of those affected by BLM's actions. Even if the lands had been within the BLM programme, the plaintiffs could not use judicial review to challenge the whole BLM programme: such challenge should be mounted in the 'office of the Department'. The court thus implied that challenges to the basic policy of an environmental agency are *political* matters and not, therefore, within the jurisdiction of the courts.

In 'Lujan II' (*Lujan* v. *Defenders of Wildlife*) the NGO sought to challenge the use of US funds to assist Indonesian irrigation projects and the rebuilding of the Aswan Dam; projects which might lead to extinction of endangered species and destruction of fragile habitats. The Supreme Court denied the NGO standing because its members had not suffered an 'injury-in-fact' apart from their special interest in the area and the species that might be affected. The fact that members of the NGO intended to revisit the project sites in the future, and might not then have the opportunity to observe endangered animals, did not demonstrate an 'imminent' injury.

In addition to 'injury in fact' US courts have added the requirements that the defendant's complained-of conduct is a *cause* of that injury, and a requirement of likelihood that the requested relief will *redress* that injury (e.g. *Chicago Steel and Pickling Co.* v. *Citizens for a Better Environment*).

In Australian and UK courts the trend has run in the opposite direction to that in the United States. Early cases restricted standing to plaintiffs who were 'adversely affected in some way to an extent greater than the public' (*Robinson* v. *Western Australian Museum* (1977)), who had suffered 'special damage' (*Australian Conservation Foundation* v. *Commonwealth* (1980)) or who were 'personally affected by the decision' (see *R.* v. *Secretary of State for the Environment, ex parte Rose Theatre Trust* [1990]). Later cases have eased these criteria in both countries. In Australia this has been achieved by taking a broad view of what it means to be a 'person aggrieved' – the relevant criterion under the Administrative Decisions (Judicial Review) Act 1977 (*North Coast Environment Council Incorporated* v. *Minister for Resources* (1994)); *Tasmanian Conservation Trust Inc.* v. *Minister For Resources and Gunns Ltd.* (1995)). In the UK the courts have taken a broad look at what it means to be a person who has 'sufficient interest in the matter to which the application relates': the statutory criterion for standing (Supreme Court Act 1981, s. 31(3)). In *R.* v. *Pollution Inspectorate ex parte Greenpeace (No. 2)* [1994]) Greenpeace UK was granted standing to challenge the operation of the nuclear waste and fuel reprocessing plant, THORP. The court took account of the fact that Greenpeace is a respectable and knowledgeable body, was interested and specialised in the matter in hand, and was representative of a section of the local population. The case of *R.* v. *Secretary of State for Foreign Affairs ex parte World*

Development Movement [1995] stressed that, when considering the issue of standing, the *merits* of the challenge are a critical factor. These include the importance of vindicating the rule of law, the importance of the issue raised, the likely absence of any other responsible challenger, the nature of the breach of duty against which relief was sought, and the role of the applicants in the area concerned. According to Sedley J in *R.* v. *Somerset County Council and ARC Southern Limited ex parte Dixon* [1997] the notion that standing in English law demands any kind of special interest is 'entirely misconceived'.

Developing countries have tended to take a very pro-environmental stance on the question of standing:

- The Supreme Court of India has taken the view that an individual or social organisation espousing the cause of the poor and the downtrodden should be permitted standing by merely writing a letter and without incurring any expenditure at all. This is based on a broad and progressive reading of Article 21 of India's Constitution to include a 'right to a clean environment' (Lau 1995).
- In *Dr. Mohiuddin Farooque* v. *Bangladesh and others* (1997) the Bangladesh Appeal Court determined that 'a person who agitates a question affecting a constitutional issue of grave importance, posing a threat to his fundamental rights which pervade and extend to the entire territory of Bangladesh' has standing.
- In *Juan Antonio Oposa and others* v. *Hon'ble Fulgencio S. Factoran and another* (1994) the Philippine Supreme Court granted standing to challenge a decision not to revoke timber felling licences on the grounds that the children who brought the action 'represent[ed] their generation as well as generations yet unborn'.

The fourth limitation is one of costs. To bring an action for judicial review in most common law jurisdictions is to run the risk of financial ruin. The general rule for the award of costs in common law states is that 'costs follow the event'. This means that a litigant may find herself liable not only for her own costs, but also for the costs of the public body whose decision is being challenged: a rather unappetising prospect. Courts have discretion to depart from the general rule in unsuccessful but meritorious cases. In the UK costs awards have been applied inconsistently, making it difficult for would-be challengers of environmentally harmful decisions to assess their potential financial risks (Carnwarth 1999).

A potentially attractive solution to the problem of costs is the provision of some form of legal aid or assistance. In the UK legal assistance is very

public benefit is missing. In *Re Grove-Grady* [1929], for instance, a dedication of property as an animal refuge was held not to be a valid charitable trust, because – according to the court – the public would obtain no tangible benefits from it. The speech of Lord Hanworth Mr, in which his lordship objects to the proposed object of the trust (the establishment of animal refuges), indicates a distinct hostility to the conservation of ecosystems and yet, paradoxically, also hints at an underlying sympathy for the preservation of endangered species:

> In this proposed refuge all creatures are to find freedom and safety from molestation or destruction by man. Whatever animals, birds or other creatures, ferae naturae, may be living therein or obtain access thereto, are to find sanctuary. The fox and the rabbit, birds of all sorts with the stoat and the weasel and rats as neighbours, and hawks and crows as spectators, are to live and enjoy themselves after their kind. The struggle for existence is to be given free play [reference to Darwin's *Origin of Species*]. It is not a sanctuary for any animals of a timid nature whose species is in danger of dying out: nor is it a sanctuary for birds which have almost left our shores and may be attracted once again by safe seclusion to nest and rear their young. No such purpose is indicated, nor indeed possible. The one characteristic of the refuge is that it is free from the molestation of man, while all the fauna within it are to be free to molest and harry one another. Such a purpose does not, in my opinion, afford any advantage to animals that are useful to mankind in particular, or any protection from cruelty to animals generally. It does not denote any elevating lesson to mankind.
>
> (*Re Grove-Grady* [1929], pp. 573–5)

The Australia case law shows a strong trend towards acceptance of charitable trusts for environmental protection. In *Halliday* v. *Institute of Foresters of Australia* (unreported), a case involving a legacy for 'the preservation, propagation and dissemination of Australian flora', YOUNG J noted that

> One starts from the proposition that in these days gifts for preservation of fauna and flora appear to be charitable, notwithstanding that in an earlier age a different view was taken: see *AG* v. *Sawtell* (1978) 2NSWLR 200 and *Brisbane City Council* v. *Attorney General Queensland* (1978) 52ALJR 599.

Similarly, in *Fellows* v. *Sarina and others* (1996) it was held that a provision of a will that certain land was not to be sold but was 'to forever remain as a wildlife sanctuary and can be used by all my relations and their friends to picnic or camp thereon' could be for a valid charitable purpose whether or not access by the public was permitted.

A second form of trust – the *public trust* – also holds considerable scope for environmental protection. A public trust exists where the state asserts ownership of property for the use and benefit of the whole public. As such, the public trust doctrine forms the basis of state action to protect areas such as the foreshore, lakes, rivers and mountains from environmental degradation (Sax 1972; von Tigerstrom 1998). A public trust could, in principle, be used to advance the ethical idea that interaction with wild nature is viewed as an essential precondition for proper human development. Courts, especially in the United States, have invoked the public trust as a measure to justify environmental protection. The public trust doctrine can be used to create a procedural duty to take a 'hard look' in environmental cases (Baker 1995). For instance, in *National Audubon Society* v. *Superior Court of Alpine County* (1983) the Supreme Court of California required that in granting any authorisation to divert the waters of Lake Mono to maintain water supply to Los Angeles,

> state courts and agencies . . . should consider the effect of such diversions upon interests protected by the public trust and attempt, so far as feasible, to avoid or minimize any harm to those interests.

Unfortunately, property rights can inhibit as well as promote environmental protection. An obvious aspect of this is that whilst property provides a degree of freedom from pollution, or non-consensual exploitation, property also legalises exploitation by its owners. Hardin's concerns about over-exploitation of commons are resolved by privatisation if, and only if, holders of private property think it worth while to engage in sustainable practices. Since many people – especially corporations – are generally interested in short-term gains, property rights may actually *encourage* environmental destruction. This feature of law extends to individual animals and plants. Plants are considered to be part of the property ('fixtures') on which they are situated. Animals, once reduced into possession (i.e. captured), are in English law the property of the landowner (Bonyhady 1987). Animals not on land (e.g. whales or dolphins caught at sea) are also considered to be the property of those who take them or who first strike them (*Swift* v. *Gifford* (1872)). The very notion of 'property' in wild animals is abhorrent to those who find animal rights arguments convincing.

In the United States, property rights can impede environmental protection by providing a right to compensation under the doctrine of 'takings'. The takings doctrine stems from the Fifth Amendment to the Constitution which stipulates that compensation shall be paid if private property is 'taken' for public use. Two apparently independent tests for the existence of a taking exist (Meier 1995). The older of the two (*Penn Central Transportation Co.* v. *City of New York* (1978)) is that in determining whether a taking has occurred the court must balance three factors:

- the character of the governmental action;
- the extent to which that action interferes with reasonable, investment-backed expectations; and
- the economic impact of the action on the claimant.

The later test (*Agins* v. *City of Tiburon* (1980)) is that a law may amount to a taking if it denies the owner an economically viable use of the land or if it does not substantially advance legitimate state interests.

Environmental legislation is particularly susceptible to challenge as a taking on these grounds. For instance, in *Whitney* v. *United States* (1991) the federal Surface Mining and Reclamation Act, made a mining company's land totally useless, and so amounted to a taking. In *Lucas* v. *South Carolina Coastal Council* (1992) the plaintiff successfully alleged a 'taking' in relation to a South Carolina statute which, for environmental reasons, prohibited construction on part of his beachfront property.

Not surprisingly *Lucas* has been criticised as an impediment to conservation legislation. However, it should be noted that the tests sometimes result in a 'no-takings' ruling (Babcock 1995). For instance, in *Stevens* v. *City of Cannon Beach* (1994) the court took the view that a law prohibiting construction on part of the plaintiff's beachfront lot was not a taking. In this case the public had customary rights of access to the beach which were incumbent on the title at the time that Stevens acquired the land (Meier 1995). Significant exceptions to the *Lucas* ruling exist which, some have argued, virtually undo the case itself (Sugameli 1999).

There is no direct equivalent to the 'takings' doctrine in other common law jurisdictions, although there are signs of its emergence in Canada and Australia (Schneiderman 1996; Sperling 1997). In Australia Section 51 (xxxi) provides that when the Commonwealth acquires property from any state or person, it must do so on just terms. However, this does not imply an automatic right to compensation for environmental legislation that merely restricts the use of land. On this matter the various Australian environmental and planning law statutes are mixed and inconsistent (Bonyhady 1992)

Although the European Convention on Human Rights contains a right to property (Article 1 to the first Protocol), the Convention permits states, without compensation, 'to enforce such laws as it deems necessary to control the use of property in accordance with the general interest' (Desgagné 1995). This includes laws for the protection of the environment. In *Fredin* v. *Sweden* (1991), for instance, the European Court of Human Rights upheld the legality of a decision to revoke a gravel extraction permit. EU law embodies the same basic position: environmental regulation is not generally an infringement of the EU's fundamental right to hold property (*R.* v. *Secretary of State for the Environment and another, ex parte Standley and others* [1999]). An exception to this last rule may occur where environmental law is so intrusive that it requires the actual destruction of property, in which case it may be that compensation must be paid. In the *Booker Aquaculture Ltd* v. *Secretary of State for Scotland* (2000) litigation, currently before the ECJ, the applicants have claimed that an order, made without compensation, requiring the company to destroy its stock of farmed fish, infringes a basic EU right to property.

The UK not only has no notion of takings, but allows for the opposing principle of 'payment for freedom to develop': property owners wishing to obtain development permission may lawfully provide environmental or other public benefits in return (*Tesco Stores Ltd.* v. *Secretary of State* [1995]).

Property law has philosophical roots that are less than environmentally benign. The common law generally takes *possession* to be the root of title to property (Epstein 1979; Rose 1985). One advantage of possession is that it is, in the main, the way that societies have traditionally operated, and there is value in historical continuity itself (Epstein 1979). However, for the sake of legal certainty, courts have generally required possession to be evidenced by *clarity* in the acts of control used. For instance, in *Pierson* v. *Post*, Post, a huntsman, was pursuing a fox along an unowned section of beach. Pierson, at the last minute, appeared out of nowhere and captured the fox. Pierson refused to give the fox back to Post, who sued for the indirect injury that he had suffered from loss of the fox. A majority of the court agreed with Pierson's position: by taking the fox into his possession he had established 'clear control' of the animal.

The common law also generally requires acts of possession to amount to some kind of *statement* of an intention to appropriate (Rose 1985, 77). Such statements can be implicit as, for example, in acts such as the erection of hedges, walls and fences, or explicit as in the placing of notices. The law of possession thus encourages enclosure, development and a generally defensive attitude towards land that works against the ethic of 'leaving be' or allowing nature to take its course. Western property law's reliance on possession as the key to ownership

> reflects the attitude that human beings are outsiders to nature. It gives the earth and its creatures to those who mark them so clearly as to transform them, so that no one else will mistake them for unsubdued nature.
>
> (Rose 1985, 88)

Besides possession, the other main classical justification for property is John Locke's theory of labour (Locke 1988, orig. 1690) which contends that property is both created and justified by the intermingling of a person's *labour* with a thing. Locke's reasoning is simple: unless labour creates property (i.e. a right to exclude others) then there will be no incentive for a person to labour: for who would sow and tend crops if anyone could come along and take them away? In the absence of the labour–property link, Locke argues, the earth will lie unutilised and people's lives will be miserable. Hargrove's historical analysis (1989) suggests that Anglo-American attitudes towards land have been heavily influenced by Lockian labour ideas which

> have encouraged landowners to behave in an antisocial manner and to claim that they have no moral obligation to the land itself, or even to the other people in the community who may be affected by what they do with their land.
>
> (Hargrove 1989)

Land-use plans, or systems of development control such as found in the English Town and Country Planning Act 1990, could, in principle, deal with this selfish and amoral attitude. Research indicates that in practice the rules of almost all systems of development control are 'bent' and subverted by market priorities (Benfield 1998). Hargrove attributes the readiness of American citizens to challenge zoning and land-use plans to the 'Lockian legacy' which encourages individuals to think of any restriction on their use of land as an infringement of a basic human right. What is often overlooked, apart from the weaknesses in Locke's theory itself, is that Locke's principles were subject to the caveat that there remained 'as much and as good' for others. In the modern world this essential precondition no longer exists.

Although law does not generally accept that labour, of itself, creates property (witness the unsuccessful hunter in *Pierson* v. *Post*) it nevertheless favours those who labour on the land against those who seek to keep land idle. For example, in an early English property case, *Salvin* v.

North Brancepeth Coal and Coke Co. [1874] Sir W. James M.R. endorsed the pro-industry position:

> If some picturesque haven opens its arms to invite the commerce of the world, it is not for this court to forbid the embrace, although the fruit of it should be the sights and smells and sounds of a common seaport and shipbuilding town, which would drive the Dryads and their masters from their ancient solitudes . . . If [the defendants] had been minded to erect smelting furnaces, forges and mills which would have utterly destroyed the beauty and amenity of the Plaintiff's ground, this court could not have interfered. A man to whom Providence has given an estate, under which there are veins of coal worth perhaps hundreds or thousands of pounds per acre must take the gift with the consequences and concomitants of the mineral wealth in which he is a participant.

The reason that the law refuses to recognise rights to a view or 'spectacle' (*Victoria Park Racing and Recreational Grounds* v. *Taylor* (1937) or a 'right to a view' (*Hunter* v. *Canary Wharf Ltd.* [1997]) is that such rights would have a stifling effect on the development of neighbouring land.

Emond (1984) has argued that property law is conceptually incapable of reflecting modern ethical attitudes towards the environment since it is founded on a philosophy of unrestrained competition between members of society which can only be resolved by (re)transforming the verb 'to own' to its etymological root 'to owe' (a duty – both to the land itself and to future generations).

Key points

- Environmental standards, if carefully designed, have considerable scope for environmental protection. Different types of standards are suitable for different forms of environmentally harmful process.
- Self-regulation and monitoring, and auditing systems, are increasingly viewed as cost-effective means of improving on direct regulation of standards.
- Property rights, environmental education, citizen enforcement and judicial review have major roles to play in a comprehensive approach to environmental management.

Further reading

Hawkins, K. (1984) *Environment and Enforcement: regulation and the social definition of pollution*, Oxford: Clarendon. Dated, but still the best exposition of the inadequacies of the softly-softly UK style of direct regulation.

De Prez, P. (2000) 'Excuses, Excuses: The Ritual Trivialisation of Environmental Prosecutions', *The Journal of Environmental Law*, vol. 12, no. 1, pp. 65–78. This provides empirical research into the in-court techniques used by illegal polluters to avoid conviction or achieve lower sentences.

Hayward, T. and O'Neill, J. (1997) *Justice, Property and the Environment: social and legal perspectives*, Aldershot: Ashgate, and Freyfogle, E.T. (1993) 'Ethics, Community, and Private Land', *Ecology Law Quarterly*, vol. 23, pp. 632–61. Both sources provide some thought-provoking material on the suitability of reliance on property-rights for environmental protection.

Discussion questions

1 Is the environment better served, overall, by rigorous direct regulation, by extension of the use of self-regulation, or by some combined approach?

2 Can businesses be trusted to engage in self-regulation for environmental protection?

3 Should compensation be payable to the owners of private property whose use of land is restricted by environmental laws?

4 An environment minister from a developing country asks you for your advice concerning the best techniques to be adopted in dealing with reduction of (a) water pollution from farms, (b) air pollution from a small number of closely situated industrial factories, and (c) damage to sites of special ecological importance. In each case outline the legal technique(s) that you would recommend be employed and justify your choices.

6 The economics of environmental law

- **Different views on the economics/environment nexus**
- **The use of charges and taxes**
- **Subsidies**
- **Transferable permit systems**

Economics is a powerful and important field of study for those interested in halting environmental degradation. This chapter explores a number of key linkages between economics, law and the environment. The first, most general, connection is that economic activity is both the *cause* of environmental harm in the form of pollution and loss of biodiversity and, paradoxically, the *means* by which resources can be generated that may be employed in addressing environmental harm. This is examined below in relation to the competing visions of economic pessimists and ecological modernists. The second linkage is between economic measures, or 'instruments', and changes in human behaviour that are favourable to environmental protection. Environmental taxes, subsidies, and tradable permit systems all have a role to play in reducing environmental harm or, at least, in allowing the ends of environmental protection to be achieved at the least cost to society. Finally, the chapter examines the extent to which the common law, in areas of property rights and the tort of nuisance, can be seen as providing an 'economically efficient' means of environmental protection.

Economic pessimism, ecological modernisation and environmental law

Economic pessimism

The planet on which we live and the biological systems which support life are, without doubt, finite. The question is: how finite? Is the massive expansion of economic and technological growth threatening Gaia – the whole living planet system (Lovelock 1979). Lovelock suggests that the earth as a whole is one massive living organism, with homoeostatic (i.e. self-regulating) properties. The basic evidence for this is quite strong: the earth's atmosphere has, since its origination, remained in a chemically unstable condition; a fact only explicable by reference to some self-organising and sustaining capacity. The important question remains, however, whether the side effects of the unique sustained economic growth of the last 150 years have already disturbed Gaia beyond its boundaries of resilience? In order to better consider that question we need to make some assumptions about the kind of super-organism that Gaia is. Figures 6.1–6 give a number of theoretical models for Gaian resilience. The models represent the earth's current ecological state as a small ball, and the underlying nature of Gaia as a curve, or series of curves (adapted from Milton 1991).

Model A (Figure 6.1), the most pessimistic assumption, envisages the earth as a sensitive delicate organism, which can be pushed into irreversible decline by quite small disturbances. If Model A is correct then the alteration of the atmosphere by increases in carbon dioxide and other greenhouse gases, along with depletion of the ozone layer and general bio-abundance, could already have pushed the earth into the first stages of a decline from which it cannot recover.

Model B (Figure 6.2), less pessimistically, accepts that the current climatic and biotic equilibrium may be easily disturbed, but that after a time Gaia will find a new more stable equilibrium point which is more robust than that which currently exists.

Model C (Figure 6.3) could be described as the most optimistic Gaian position: the earth is strong and will always return to the current climate and biotic conditions even if strong disturbances take place.

Model D (Figure 6.4) modifies this by assuming the existence of fragile boundaries which, once exceeded, cannot be recovered.

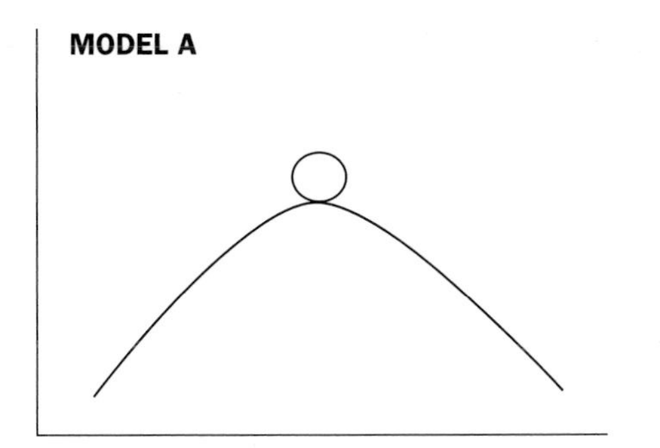

Figure 6.1 ***The earth's biosystems are very fragile***

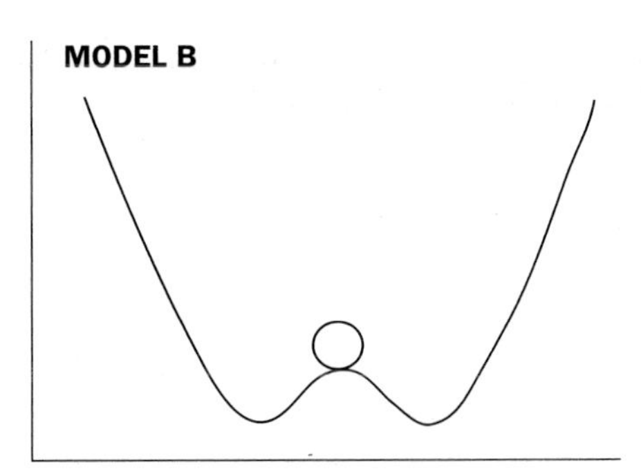

Figure 6.2 ***The earth's biosystems are fragile within robust boundaries***

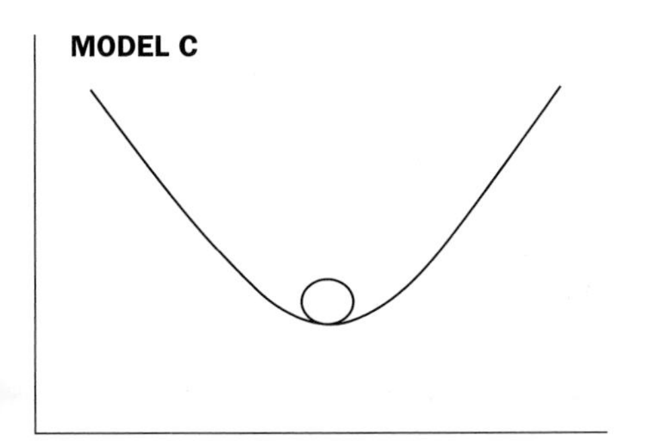

Figure 6.3 ***The earth's biosystems are very robust***

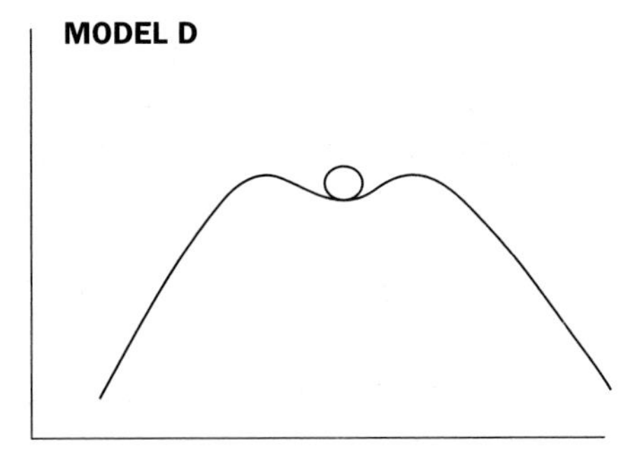

Figure 6.4 ***The earth's biosystems are robust within fragile limits***

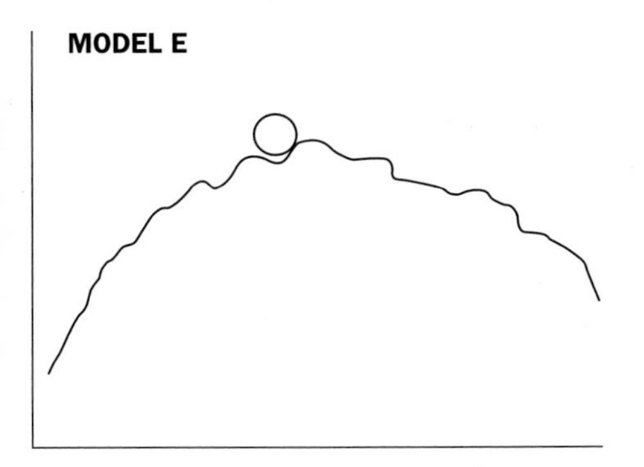

Figure 6.5 ***The earth's biosystems have multiple localised points of robustness, but are essentially fragile***

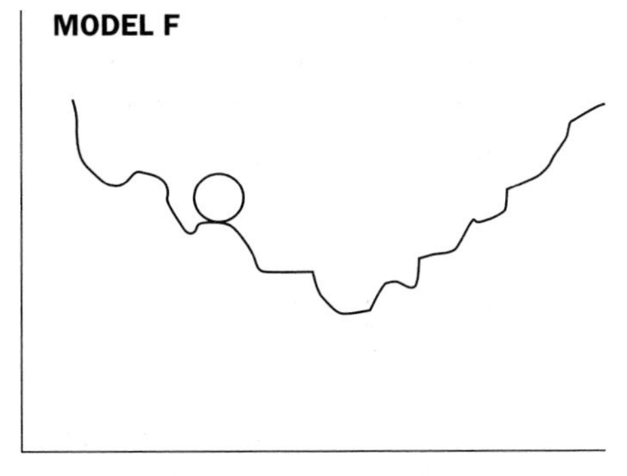

Figure 6.6 ***The earth's biosystems have multiple localised points of fragility, but are essentially robust***

Models E and F assume the existence of multiple combinations of fragility and robustness. In E (Figure 6.5) the basic pattern is fragility with odd spots of robustness. In F (Figure 6.6) it is the other way round.

Now the problem for society in assessing economic acitivity is that whilst we know that all economic activity causes climatic and biotic alteration, we do not know the basic characteristics of the planet on which this activity takes place. Palaeoclimatology (i.e. the studies of climates of past geological time zones) reveals that the earth's atmosphere and climate have undergone very considerable change over many thousands of million years. This might be considered evidence that Model A, at least, is not correct. It might also be taken to indicate a general robustness (i.e. Models B, C or F). On the other hand our neighbouring planet, Mars, provides us with evidence that atmospheres, even those which at one time supported ice or water, can be lost. We may not, therefore, confidently assume full robustness.

Palaeoclimatological evidence of past climatic changes (Harvey 2000, 4) indicates that some version of Models E or F most closely approximate Gaia; that is, that there are areas of stability and fragility within an essentially robust or fragile system. Assuming this to be correct then it is probable that the consequence of human activities to date will be a shift away from the current climatic/biotic conditions to a new point of balance. Just how far removed from our present world is hard to judge. Current estimates are that a doubling of CO_2 levels will result in a global mean warming of between 2.1°C and 4.8°C (Harvey 2000, ch. 9). Before carbon fuel use began the earth's atmosphere had only 580 million tonnes. By 1991 that had increased to 750 billion tonnes, increasing at 6 billion tonnes per year. Doubling is thus predicted to occur within this century, around 2060 (Leggett 1999) and temperatures could eventually rise by up to 7°C by 2400 (Harvey 2000, 5). The last time that the earth had carbon dioxide levels as high as those now present in the atmosphere was during the Eocene period. In the Eocene there were no ice-caps and the global climate was considerably hotter than currently exists; for example, the area that is now London had an average temperature of 25°C (rather than the current 10°C) and consisted of huge areas of mangrove swamp. It is possible that things might be made considerably worse for biotic life by the damage to the ozone layer, which may adversely affect not only large fauna and flora but also phytoplankton and microbes which perform an important role in the homoeostatic regulation of atmospheric processes. It is probable, therefore, that the new point of stability will not be as favourable for human existence as the last.

Perrins (1994) argues that the resilience of ecosystems is related to the *functional diversity* of such ecosystems; that is, the 'range of responses to environmental change, including the space and time scales over which organisms are a major threat to ecosystem resilience'. Functional diversity is itself a product of biodiversity. Most observers agree that biodiversity is being lost at an unprecedented rate (e.g. Wilson 1992, although cf. Myers and Simon 1994), which suggests that ecological resilience may also be in decline. The prescription for this ailment is, Perrins suggests, adherence to sustainable lifestyles and, implicitly, restriction of technologies to types and output levels which do not threaten resilience.

In assessing this, the most general of correlations between economics and the environment, the important question is whether on the whole, and in the longer term, economic activity will eventually push Gaia beyond its boundaries or, at least, beyond those boundaries that are of relevance to human life on earth. In this general debate there are those who take the position that economic growth and environmental sustainability are, ultimately, incompatible. This view we can term 'economic pessimism'. In contrast there are those who view economic activity as the way forward to a brighter future in which environmental standards rise. This more recent school of thought is often described as 'ecological modernisation'.

One of the strands of economic pessimism is the well-known 'limits to growth' argument. Limits to growth theories are commonly traced back to the views of Malthus (1970, orig. 1798) who maintained that population growth (being exponential) would be limited by growth in agricultural productivity (being linear). It is generally accepted that, on the population question, Malthusian predictions have not been accurate due to unforeseen advances in agricultural productivity and due to reductions in birth rate that occur as a country becomes more affluent – a process known as 'demographic transition'. The fact that a purely agricultural limit to societal growth has not yet been encountered does not, of course, mean that no limits to economic growth exist. More recently Meadows *et al.* (1974, 1992) have used computer modelling to seek to demonstrate that physical limits to economic growth are present despite the advances of technology. The precise outcomes of these detailed modelling exercises have been criticised but the basic conclusion, that we are in danger of exceeding the planet's bio-systemic limits, seems sound.

Those within the field of ecological economics working with entropy theory (Georgesçu-Roegen 1971; Beard 1999) have pointed out that economic activities take place within the restraints and parameters set by natural biological and physical systems. These natural systems are subject to the laws of thermodynamics. The first law of thermodynamics states that matter cannot be created or destroyed. This has strong implications for rosy-tinted hopes of 'disposing' of waste or 'destroying' harmful substances: matter can only be redistributed or changed into different forms. Certain human-generated substances are both harmful and stable (e.g. CFCs and some pesticides) and, therefore, difficult to regulate with an appropriate degree of safety.

The second law of thermodynamics states that all processes are subject to the 'entropy principle'. Entropy is a measure of unavailable energy. Physical processes turn concentrated or usable energy or resources into dispersed and unusable energy or resources. For example, when fossil fuels are burned energy is dissipated into the atmosphere, and the carbon is released as CO_2. Thus a low entropy source (coal deposit) becomes a high entropy end product (ashes, CO_2 and waste heat). Low entropy can be gained from only two sources: existing planetary resources and, externally, from sunlight which is captured by photosynthesising plants, tidal effects from the interaction of the earth, sun and moon. Industrial processes, which form a large part of the economic development that world institutions seek to promote, transform low entropy into high entropy at a high rate. Technology can, to some extent, mitigate this effect, but there is no guarantee that technology will continue to be able to do so indefinitely. At the same time the low entropy end products (waste) have to find somewhere to go. The world only has limited 'sinks' in which low entropy products can be assimilated. Once these are full low entropy products begin to interfere with biosystems. As biosystems deteriorate the source of low entropy products (e.g. food) begins to decline and the assimilative capacity deteriorates further.

A closely related issue is whether, even if we could continually pursue economic growth, this would be desirable. There are significant social, political and environmental costs to the 'great western project' (Mishan 1993). Social and political effects are felt not only within states, whose populations are subject to increasing wealth–poverty gaps, but also between states. It has often been argued (e.g. Trainer 1985) that affluence, at the levels currently experienced in western developed states, cannot conceivably be spread and sustained at similar levels throughout the whole world. Furthermore, poverty in developing countries is partly caused, and certainly exacerbated, by the economic relations that exist between such states and those of the developed world.

The answer to entropy, and to the limits to growth arguments, is often considered to be the 'steady state' economy (Mill 1891; Daly 1992), one definition of which is:

> an economy with constant stocks of people and artifacts maintained at some desired, sufficient levels by low rates of maintenance 'throughput'.
>
> (Daly 1992, 17)

Although the rapid decline and ecological catastrophe that some radical pessimists (e.g. Goldsmith 1972) predicted has not *yet* emerged, the steady state prescription still carries considerable force. Institutions that could cater for a non-growth world might include controversial measures such as tradable birth licences for women, depletion quotas for all non-renewable resources, and maximum limits on income and capital accumulation (Daly 1992).

Ecological modernisation

The view that continuous economic growth and environmental protection are, at root, incompatible is refuted by those belonging to the school of 'ecological modernisation'. Economic growth generates resources over and above those required by society for its basic needs, resources which can be redirected to ameliorating or reversing the negative side-effects of the processes by which economic growth takes place. This implies rejection of the steady state model:

> Nothing could cut more dangerously into the resources that must be devoted to the Great Cleanup than an attempt to limit resources available for consumption. By ignoring the prohibitionist impulse and allowing everybody to have more, we shall also have more resources to do the environmental job.
>
> (Walrich, cited by Daly 1992, 100)

The ecological modernist argument is, in essence, that the correlation between economic activity and environmental quality can be described by a U-shaped curve (Figure 6.7).

The early stages of economic growth produce a net loss of environmental quality but once society reaches a certain per capita income the relationship reverses and further growth produces improvement in environmental quality. The thinking behind the model is that people in poor countries cannot afford to, and are not inclined to, save the planet:

nationality, membership of a particular social group or political opinion'. In *Guo Chun Di* v. *Carroll* (1994) the petitioner, a 28-year-old Chinese citizen, fled to the USA after being ordered to report for a sterilisation operation. Lacking the necessary documentation the petitioner was charged with entering the USA in violation of federal law and, in response, raised the argument that he sought legitimate asylum. The key issue for the court was whether, given the threat of compulsory sterilisation, Chun Di had a well-grounded fear of persecution. The court's conclusion was that the ambit of 'political opinion' includes an individual's views regarding procreation because the right to bear children is a basic human right.

If one looks at the other side of this divergence of view, one finds evidence that law often favours the ecological modernist position. Environmental laws rarely hinder the principal objective of free trade; indeed trade is itself taken to be a prerequisite for economic growth. The Danish Bottles case (*Commission* v. *Denmark* [1988]) illustrates the difficulty of striking a happy balance between economic growth and environmental protection. This case arose from a dispute between the European Commission and Denmark over a Danish scheme requiring manufacturers to market beer and soft drinks only in reusable containers. Under the Danish scheme drink could only be sold in containers approved by the national environmental agency. The Danish agency could refuse approval of new kinds of container, especially if it considered that a container was not technically suitable for a system for returning containers or that the return system envisaged would not ensure that a sufficient proportion of containers were actually reused. Denmark modified the system to allow small quantities of non-approved containers to be used, but the European Commission nevertheless brought Denmark before the European Court of Justice, alleging that the mandatory reuse system infringed Article 30 of the Treaty prohibiting 'quantitative restrictions on imports and all measures having equivalent effect'.

In its judgment the ECJ recognised that rules for environmental protection could amount to a legitimate interference with free trade, so long as these rules were non-discriminatory between Member States, and were proportionate to their objectives. However, it also took the view that a state's environmental objectives must themselves be moderate. The Danish system, in seeking to achieve full or nearly full recovery of drinks containers, was excessive in its objectives and could not, therefore, be considered as a proportionate measure. A system requiring deposit and return would have been permissible, but the requirement for type approval of bottles went too far. One interpretation of the ECJ's conclusion is that they were, in effect, denying EU Member States the right to choose their own level of environmental protection (Krämer 1993, 103). Another way of looking at the decision is to conclude that only small interferences with economic activity could be considered to be acceptable in the name of environmental protection; more significant restrictions would be considered excessive restrictions on free trade.

Economic/market instruments

One of the most significant links between economics and environmental protection is the development of economic instruments to perform this task. Realisation that traditional 'command-and-control' regulation is beset with certain problems, as discussed above, has led to a general view that the use of economic or 'market' instruments is, where possible, preferable (see, e.g., Rio Declaration Principle 16; the 5th EU Action Plan on the Environment, and the UK document *This Common Inheritance*, Second Year Report 1992 Cm 2086). Although there is a wide range of instruments and mechanisms which exist that could be broadly described as 'economic' (see Bándi 1996, for an overview), this chapter focuses on three central models:

- charges or taxes;
- subsidies; and
- transferable permit systems.

Much of the discussion of these models revolves around two key concepts, which deserve explanation in advance. The first is the economic notion of *efficiency* or, as it is also known, an *efficient outcome*. To achieve efficiency or an efficient outcome is to deliver some policy goal (indeed *any* policy goal) at the least overall cost to society. Importantly, the concept of efficiency is quite independent from the question of values or desirability. One can have highly efficient systems that are to be rejected as ethically bad or wrong (e.g. efficient genocide). Also note that efficiency has no necessary relation to the concept of effectiveness, the latter being about success in delivering ends rather than achieving ends at least cost.

The second important economic concept is that of *optimality*. Economists speak of a policy or technique being *optimal* when it results in the greatest level of social welfare possible. Optimality does not necessarily imply efficiency, since one can employ devices with varying degrees of efficiency to achieve optimal outcomes. Nor does the economic concept of optimality imply, in broad sense, the overall best thing. Maximising human welfare, even over long periods of time, may require serious damage to plants, animals, species and ecosystems. Optimality, in this weak sense, may not be what is demanded by environmental ethics (see Chapter 8).

The two concepts can be concretised by considering the matter of pollution. An optimal level of pollution is one that results in the greatest overall level of social welfare for the longest period of time. This pollution level is certainly not zero, since welfare requires human activity and the second law of thermodynamics requires all activities to result in some waste. On the other hand it is certainly not as high as the level of pollution that would ensue if no controls were imposed. So it must lie somewhere between these two extremes. In fact, economists demonstrate that optimal levels of pollution are those which arise from the abatement which firms voluntarily undertake when the external costs of their pollution are reflected back to them; that is, when their marginal pollution costs are internalised (Pearce and Turner 1990).

Efficiency is another matter. For any level of pollution that society has deemed desirable there will be a range of methods for effecting the reductions in pollution output necessary to achieve that level. The methods with the least costs are 'efficient', even though the overall pollution target itself may or may not be optimal.

Charges

Charges and taxes have major strengths. Charges and taxes can be *efficient* in the sense discussed above; that is, they can allow an industry (or society) as a whole to achieve a politically determined amount of pollution abatement more cheaply than a 'command-and-control' approach. Why is this? The key lies in the fact that polluters are heterogeneous (i.e. they have different marginal cost curves). This entails that some can abate pollution more cheaply than others or, to put it another way, achieve more abatement for a given sum expended. Firms or individuals that can abate most cheaply thus have an incentive to abate the most, whilst those who would find it least cheap are able to abate less. In this way the same overall level of pollution abatement can be delivered at a *lower total cost*.

Quite independently from the advantage of cost-effectiveness, charges and taxes can also be *efficient*. More precisely this means that they can, if set at the right level, achieve the 'optimal' level of pollution. This is the level of pollution corresponding to an amount of abatement for which the total of social and private costs is minimised in comparison to its total benefits (Pearce and Turner 1990).

An important advantage of charges and taxes is that they obviate the regulator's need to obtain information from individual polluters in order to achieve a given level of pollution at minimal cost. Each firm self-monitors its own costs and so will spend money on abatement when this is cost-effective, but will pollute when paying the tax is less expensive than abatement. Management insight is thus given full incentive to innovate and discover new production methods that generate lower levels of pollution.

Taxes and charges can be used to promote higher technological standards. In Germany, charges levied by the *Länder* (states) on all direct discharges to watercourses are based on the number of 'units' of pollution released, which in turn reflects the volume and content of effluent. Polluters who meet minimum standards of waste water treatment pay a lower charge rate than those who do not. In Sweden differential taxes have been used to speed up the change to cars equipped with catalytic converters. Cars not equipped with a converter were taxed whilst cars fitted with this technology were exempt from the tax (Tietenberg 1990, 20).

Finally, the funds generated by taxes and charges can be deployed for the benefit of the environment or those persons dependent upon it. In France, funds from environmental charges are used to subsidise the deployment of clean technologies. In Japan air pollution charges are used to compensate air pollution victims.

In Britain the use of environmental charges is relatively undeveloped. Polluters pay trade effluent charges for discharges into public sewers based on a nationally agreed 'Mogden' formula (including parameters for Chemical Oxygen Demand (COD) and suspended solids). These charges, however, merely reflect the costs of reception and carriage of the pollution and the costs of primary treatment (Royal Commission on Environmental Pollution 1994, 151). In 1990 the National Rivers Authority introduced a charges scheme for recovery of the costs of determining, issuing and monitoring discharge consents to surface waters. Neither scheme is designed to be an environmental tax, but both place financial burdens on polluters and, as such, provide incentives to reduce pollution (Hughes

1991). The Royal Commission on Environmental Pollution (1994, 151) concluded that a UK incentive charging scheme could improve cost-effectiveness, encourage new technology, reinforce the 'polluter pays principle', send a strong message to reinforce public attitudes towards pollution, reinforce regulation, and act as a revenue source. The UK government has not, however, taken the same view and has recently decided against the introduction of such a measure (ENDS 1999g).

Although it was a 'slow starter' in the use of environmental taxes, the UK is now adopting a wide range of such instruments. UK environmental taxes now include the landfill tax on waste, motoring taxes (differentiated road fund tax and fuel duties), an impending aggregates tax and an impending climate change levy.

The landfill tax, imposed by the Finance Act 1996, is a tax levied at two rates (inert waste £3 per ton, non-inert waste £10 per ton) on wastes going to landfill sites. A small proportion of the finance raised by the tax can effectively be diverted into environmental projects through the availability of tax credits where the operator of the landfill site makes a donation to an approved environmental body. There are also exemptions to the tax designed to encourage environmentally desirable outcomes. These include temporary storage pending reuse, landfill of mine and quarry waste, and landfill resulting from the reclamation of contaminated land.

Fuel duties have been increased at a rate exceeding inflation in the UK since 1994. Fuel tax makes up about 80 per cent of the cost of fuel in the UK: a much higher figure than in most other countries. The contribution of these high tax rates to an overall high cost of fuel has recently led to political unrest and direct action. Fuel tax differentials have also been used within fuel sales to select for cleaner fuels. Duty on leaded petrol was increased more quickly than duty for unleaded fuel (although leaded fuel is now almost completely prohibited). Tax on high octane unleaded fuel is higher than that for normal unleaded fuel to reflect the higher benzene content of the former. A two pence per litre tax differential between ordinary diesel and ultra-low-sulphur diesel (ULSD), reflecting the lower particulate pollution from the latter, has resulted in the ULSD market share rising from 3 per cent to 43 per cent in around six months. Vehicle excise duty (VED) is now also differentiated. Cars with small engines (up to 1100 cc) pay around two-thirds of the tax paid by other vehicles. VED for lorries ranges significantly from nearly £6,000 for poor environmental performers (40-tonne, five-axle lorries) to only £1,000 for low-emission lorries. Major reforms of company car taxation, reflecting the impact of company car usage, are currently being introduced (ENDS 1999h).

The climate change levy is a tax imposed on electricity use, designed to assist the UK in achieving its targets to reduce greenhouse gas emissions. The UK's legal obligation under the Kyoto Protocol is to reduce greenhouse gas emissions by 12.5 per cent by 2008–12 and it has issued a domestic pledge to reduce CO_2 emissions alone by 20 per cent from a 1990 baseline by 2010. The electricity tax has a modest aim of reducing business CO_2 emissions by 2 per cent and contains relief measures for major energy users (ENDS 1999f). The impending tax has prompted several industrial sectors to engage in negotiation of voluntary energy efficiency agreements (ENDS 1999i).

Charges and taxes have some theoretical difficulties (Baumol and Oates 1975):

1 In order to use taxes to achieve an *optimal* level of pollution, a legislator requires information both about industry's marginal costs and the marginal costs of the resultant pollution damage. A 1974 study by the UK government into charges on the River Tees found that information about industry costs was only forthcoming with the goodwill of the companies concerned, which could not be assumed to exist in a 'real life' situation (Royal Commission on Environmental Pollution 1994, 145). Difficulty in the calculation of industry costs is compounded by the appearance of 'economies of scale' for those who invest most heavily in pollution abatement (Pezzey 1988). Computation of social and environmental costs involves enormous difficulties and probably is not practical (Findley and Farber 1992, 157). This drawback does not, of course, prevent charges and taxes from being *efficient* (see p. 181).
2 The most efficient charge schemes, paradoxically, are likely to be the most complex and inequitable. The 1974 Tees study, for instance, found that large variation in unit charges between pollutants and between locations along the length of the river would be required to achieve optimal results.
3 Charges and taxes may not always result in the expected changes to polluter behaviour. This may be due to internal barriers, information deficits (Royal Commission on Environmental Pollution 1994, 152), or simply non-elasticity in the targeted activity (e.g. petrol taxes need to be very high to affect motoring habits).
4 Charges and taxes may be no better than incrementally tightened direct regulation in their ability to promote innovation or changes in technology (Rose-Ackerman 1977).
5 Charges and taxes must be used with care since they may have adverse

social as well as positive environmental implications. A water-use charge, for instance, may discriminate against the poor and infirm since these groups have higher than average water needs (Legge 1992).

6 Charges and taxes may fail to distinguish between clean and non-clean technologies (e.g. road taxes based on vehicle engine size give no incentive to produce clean-burn engines).
7 Unilaterally applied charges and taxes can disadvantage domestic industries, thus resulting in a transfer of production to other less well regulated states (Helm and Pearce 1990). This can be overcome by reduction in non-environmentally related taxes (e.g. the UK Landfill Tax is offset by a 0.2 percentage point cut in National Insurance (i.e. social welfare tax) contributions).

Subsidies

Subsidies – which can include direct aid, grants, soft loans and financial guarantees (Bándi 1996) – involve a transfer of capital from the state to private individuals or other states in return for adoption of, or abstention from, certain practices or technologies. From the point of view of behaviour modification, subsidies operate in a manner directly opposite to taxes or charges. They are, however, theoretically just as effective as a means of minimising overall costs: a polluter will invest in pollution abatement or conservation up to, but no further than, the point at which the costs of abatement or conservation are just equal to the level of the subsidy received. Some operators will be able to do this more cheaply than others and, therefore, the overall cost of conservation or abatement will be reduced (Pearce and Turner 1990). Subsidies have the potential to speed up investment in abatement and 'clean' technologies so that, in the future, environmental harm is lessened. Farmers and foresters can be given payments in exchange for agreeing to engage in certain environmentally friendly practices. In the EU, Regulation 797/85 allows Member States to provide grant aid to farmers in areas of high conservation value, and that are at risk of environmental degradation, in return for less intensive farming practices.

A further use for subsidies is in correcting market imperfections that would otherwise result in too much of certain types of environmental harm. This point derives from the so-called 'Theory of Second Best'. If the market is incapable of fully reflecting the marginal social cost of a given activity in the price charged for the goods or services produced by that activity, then overconsumption of these services will probably ensue.

One means of addressing this, given sufficient elasticity in consumer demand, is to subsidise a competing activity. Payment made to organic or non-intensive farmers, who face competition from traditional intensive farms, is an example of a 'second best' subsidy. It is difficult to devise policies which fully capture the environmental costs of intensive farming; reducing the costs of non-intensive farming through a subsidy helps to redress the imbalance (i.e. artificially low demand for the goods produced by the latter).

Subsidies are not always helpful to the environment. The most common use of subsidies at the state level is in the form of payments to farmers to provide, in effect, a level of social support when market prices are low. Agricultural subsidies, such as found in the EC's Common Agricultural Policy, can be very destructive owing to their effect of encouraging needless levels of intensive farming practices. A specific example of this undesirable effect is the Directive on Mountain and Hill Farming in Less Favoured Areas (75/268/EEC) which allows for special aid to be paid to farmers in 'areas in danger of depopulation and where the conservation of the countryside is necessary'. Ironically, payment to UK farmers based on an annual 'headage payment' for livestock, and above average levels of grants for investment in land 'improvement', have resulted in ecological overloading of many upland areas.

Subsidies in general have considerable potential to distort the conditions of international trade and upset the level economic playing field. Subsidy recipients can offer goods at lower prices than non-recipients. Subsidies can attract new entrants into the very economic activities that they seek to reduce (Pearce and Turner 1990, 107–9). Thus many legal systems restrict their application. EU law, for instance, generally prohibits subsidies by Article 92 of the EC Treaty. The OECD recommends limiting subsidies to measures that do not 'create significant distortions in international trade and investment' and to investments in technologies for the abatement of *existing* pollution (Beckerman 1975, 37). The EU Commission, which has the power to approve state subsidies in certain cases (Article 93), has as a matter of policy restricted payments to installation of anti-pollution technology for existing businesses. Subsidies may also be seen as unethical – first, because they reward rather than punish polluters and, second, because they generate inequity between subsidy recipients and non-recipients.

Tradable pollution permits

Tradable pollution permits allow pollution rights to be exchanged with relative freedom, between polluters. Their main advantage is in allowing a *given* level of environmental improvement to be attained at the *least* overall cost to the industry concerned. Polluters that can abate more cheaply can abate more than is required, and then sell the 'extra' reduction that they achieve to those firms that find abatement of pollution more expensive.

Tradable permits have been used for some time in the United States under the Clean Air Act programme (Tietenberg 1990). The Act requires the EPA to set ambient environmental standards (National Ambient Air Quality Standards – NAAQS) for a range of pollutants (generally, see Findley and Farber 1992, ch. 3). Emission standards are then imposed on a large range of pollutant points (stacks, vents, storage tanks, etc.) in order to reach the prescribed ambient standards. The standards reflect the amount of pollution that should be produced if a technology, selected by the regulator, is installed at each point. A polluter who can achieve more than the legally prescribed pollution abatement for a given discharge point may apply to the EPA for an Emission Reduction Credit (ERC). An ERC can be utilised by the same polluter to justify higher than permissible emissions at a different point, or sold to another polluter who presumably finds it cheaper to purchase the ERC than to reduce pollution.

The EPA has produced a series of policies specifying how these ERCs can be deployed and traded. Section 173 of the Clean Air Act required any new or modified stationary sources in a non-attainment area to show that the total of the emissions with the new polluter are *less* than the total from existing sources. In other words allowable emissions of each pollutant must be reduced. New sources can operate if, and only if, they first secure ERCs greater than the amount of pollution that they wish to produce. This policy balances the need for continued economic growth and progress towards lower pollution.

The EPA's *bubble policy* treats multiple existing emission points as effectively enclosed in a bubble. Only total emissions from the bubble are regulated, allowing a firm the freedom to switch emissions between specific points within the bubble. *Netting* allows ERCs to be taken into account when calculating whether existing polluters are required to undergo a source review process. *Emissions banking* allows firms to store ERCs for subsequent use within the programme or for later sale to other polluters.

The ERC trading programme has had significant successes. By the late 1980s around 12,000 ERC trades had taken place, saving over $10–500 billion in costs and increasing the level of compliance with Clean Air Act standards (Tietenberg 1990). It has also hit some problems (Dudek and Palmisano 1988; Hahn and Hester 1989). Actually generating ERCs has not been easy due to ambiguity over the baseline level which firms must exceed, or inadequate emissions data. Firms have been generally reluctant to strive to create ERCs, or to bank ERCs once created, due to uncertainty over the permanence of these property rights: they are subject to effective 'confiscation' if emission control requirements are raised by the EPA in the future. ERCs created by existing firms are sold to new entrants in non-attainment areas: thus creating new wealth for existing firms *vis-à-vis* new firms. This may have acted as a disincentive to innovation in these areas. Trades have been fewer than might be expected due to some of the EPA's own trading rules: the EPA requires new sources to meet stringent New Source Performance Standards, rather than allowing such sources to choose a lower standard and then buy extra ERCs. The EPA also requires expensive dispersion modelling for trades involving non-uniformly mixed pollutants. Some environmentalists have opposed the whole emissions trading scheme on the basis that it compounds the deficit between legal standards and attainment, and that it is, in any event, unethical since it entertains sales in clean air – a basic human and environmental right.

Tradable pollution permits have not yet been introduced in the UK, although the government is currently investigating the use of a pilot trading system as part of its obligations to reduce greenhouse gas emissions (ENDS 1999g).

The common law and economic efficiency in environmental protection

In this section we examine the efficiency of common law environmental controls. Efficiency, in this context, refers to obtaining environmental improvements at least overall social cost. More particularly, we examine two aspects of efficiency. The first is the observation that property rights – themselves a product of the common law – may be used to arrive at optimal levels of pollution; optimal, that is, judged from a purely human welfare perspective. This insight is usually attributed to the economist Ronald Coase who, in a seminal paper (1960), observed that, in the absence of high transaction costs, any clear allocation of property rights should lead to negotiated, hence optimal, pollution levels.

The second issue concerns the idea that common law liability rules, such as those that apply in the law of nuisance and negligence, maximise overall social wealth. If this is true, as argued by the American judge and economic theorist Posner, then the common law may be viewed as contributing in overall terms to the economic growth that, according to ecological modernists, is necessary for improvements in environmental quality.

Coase and property rights

It will be recalled (see p. 7) that in his classic analysis of environmental problems Garret Hardin identifies the innate tendency of humans to overexploit common resources. The solutions, he suggests, are 'mutual coercion mutually agreed upon' or an increase in private property. In the latter case it is expected that the self-interested owner will not engage in overutilisation of his own property. However, even this is no guarantee against use of the land in such a way as to impose costs on others. Take the classic 'pasture scenario'. Let us assume that the common is divided up into two halves, with one owner for each side, one with pigs and one with cows. Neither owner would, in the absence of regulation, be concerned if pollution (e.g. slurry) caused by the animals on his side migrated over to the other person's land. Such an occurrence is referred to by economists as a 'negative externality' (Pearce and Turner 1990, 61). As a rule, negative externalities allow the costs of production to be avoided by producers, and shifted to others in society. When this occurs the producers are, in effect, receiving a subsidy from those who bear the cost of the externality. In our scenario one farmer would be effectively subsidised by the other, who would have to bear the cost of dealing with the extra pollution load. In turn this would lead to artificially low costs for, and overproduction of either pigs or cows, which would not be 'efficient'.

Fortunately, property rights can, in principle, remedy this situation. In order to see how this can occur we must first distinguish property rights from liability rules (Calabresi and Melamed 1972; Landes and Posner 1987, 29). If A has a property right then the courts will prevent an invasion of that property by issuing an injunction to prevent it: they will not simply order compensation to be paid. If A merely has the benefit of a liability rule then an invasion of his right will simply lead to payment of damages. Assuming that both landowners have a property right to be free from invasion by polluting substances then it is clear that if either wishes to allow pollution to spread to the other's patch, then some agreement, probably involving payment of compensation, must take place.

The above scenario has been theoretically explored by the economist Ronald Coase (1960). Professor Coase's insight was that *any* clear allocation of property rights will, given low transaction costs, result in an efficient outcome (i.e. one in which human welfare is maximised). It matters not whether the property right is given to the polluter or the victim. If the law gives a property right allowing *freedom from pollution* then a hypothetical polluter will bargain with a victim of that pollution, the negotiated outcome being a level of pollution representing the point at which value, hence welfare, is maximised. The same outcome will result even if the property rights are reversed (i.e. the law gives a property right *to pollute*). In the latter case the hypothetical victim will negotiate with the polluter until, again, a compromise is reached and pollution is reduced to a level at which human value-satisfaction is maximised.

It will be noted that the Coase theorem makes several assumptions. The first is that property rights in a given resource can, in principle, be clearly allocated. This may not, in practice, be achievable. In the law of nuisance it can be observed that the 'right' to pollute is, generally speaking, 'fuzzy' rather than clear cut. A landowner has the right to subject his land to 'reasonable use', which neighbouring landowners must put up with. What, precisely, amounts to 'reasonable' very much depends on the locality, and the duration, the intensity and the type of activity concerned. Nuisance also requires that the pollution damage concerned should have been 'reasonably foreseeable' at the time of the activity in question. This is often difficult to achieve. Negotiations between those who have the property right not to be seriously polluted and the polluters will, in all probability, be dogged by other factors: difficulty in identifying the pollution, difficulty in identifying the source of the pollution, 'free rider' problems, and so on. In consequence, for most types of pollution, common law systems provide a relatively poor forum for the operation of the Coase theorem.

The least-cost abater

Another way in which the common law may seek to achieve efficiency in relation to the environment is by requiring the party who can most easily or cheaply avoid pollution to take such action. A number of aspects of the law of nuisance demonstrate this quality (Ogus and Richardson 1977). For instance, in certain countries or states (e.g. most US states, but not in the UK – see *Sturges* v. *Bridgman* (1879)) it is a defence to an action of nuisance to say that the claimant 'came to the nuisance'. The economic sense in this rule is that it is usually cheaper for an incoming party to choose a site away from an established polluter than it is for the existing polluter to pack up and move elsewhere. A further aspect of this rule is that nuisance applies standards for pollution according to the nature of the locality (*St Helens Smelting* v. *Tipping* (1865)); lower standards apply in an already polluted zone. Clearly, it is less expensive for an incomer to a neighbourhood to choose a different neighbourhood than for a multitude of existing operators to change their practices.

According to the American professor and judge Richard Posner common law liability rules, including those concerning the environment, operate so as to restrict damage to economically justifiable levels. This takes place, Posner (1972) argues, because judges apply, either consciously or unconsciously, the so-called 'Learned Hand' test, named after Judge Learned Hand who first enunciated the rule in *United States* v. *Carroll Towing Co*. (1947). The Learned Hand test states that liability for careless acts ought only to be imposed where the magnitude of the injury, multiplied by the probability of its occurrence, is greater than the cost of precautions which could have been taken to prevent the accident. The theory can be illustrated as follows. Imagine a factory that emits water pollution causing £100 of environmental damage. If the cost of abating that damage would have been £200, then it makes little economic sense to fix the polluter with liability. If, on the other hand the cost of abatement was only £50 then one can argue that the factory should have taken the necessary precautions. If the cost was £150 but the probability of the pollution damage occurring was only 50 per cent then, again, it would not have made sense for the factory to spend the money on pollution abatement. Thus the Learned Hand test can be seen as an extension of the 'least-cost abater' formula to take into account harm of the probabilistic nature of environmental harm.

This concludes the discussion of economics, law and the environment. Within the confines of this chapter it has only been possible to scratch the surface of a large body of literature and ideas on how economics and law can together provide for better environmental protection. The reader should be cautioned that the economists' solutions and analyses are often preoccupied with notions of efficiency, and may fail to take into account other values such as justice, political ideals, or the growing body of environmental ethics that we will discuss in Chapter 8. They are, nevertheless, of considerable importance in the overall fight against decreases in environmental quality.

Key points

- The overall role of economic growth in environmental protection is disputed: pessimists view it as a negative factor, modernists view it as an essential prerequisite for effective environmental policies.
- Charges (almost always taxes) enhance efficiency, and can – in principle at least – contribute to achieving optimal levels of pollution. They have, however, quite a few practical drawbacks.
- Tradable permits are also useful mechanisms for achieving efficient pollution abatement. These are administratively more complex than taxes but have the advantage of relative certainty concerning the resultant pollution levels.
- The common law can be analysed as having elements that contribute to economic efficiency and/or optimality, especially its creation of Coasian bargaining positions, and rules that impose the costs of prevention or liability on the least-cost abater.

Further reading

Baker, K.K. (1995) 'Consorting With Forests: Rethinking Our Relationship to Natural Resources and How We Should Value Their Loss', *Ecology Law Quarterly*, vol. 22, pp. 677–727. This paper contains an excellent study of the question of valuation of environmental damage.

More, J.D. (1991) 'A Review of Major Provisions: Clean Air Act Allowance Trading', *Environmental Law*, vol. 21, pp. 20–1.

Sagoff, M. (1988) *The Economy of the Earth: Philosophy, Law and the Environment*, Cambridge: Cambridge University Press. This book by Mark Sagoff is still probably the best read for those wishing to gain a critical understanding of the role of economics in environmental law and policy.

Discussion questions

1. Do you agree that, perhaps paradoxically, economic growth is likely to provide the key to safeguarding ecological systems?
2. To what extent is the use of economic instruments such as taxes and tradable permits *ethically* appropriate?
3. Does the use of environmental taxes/tradable pollution permits diminish public influence in determining the kind and level of environmental protection that society shall have?
4. Does it matter whether laws for the protection of the environment are economically efficient?

7 Science, law and the environment

- Scientific methodology in relation to environmental protection
- Dealing with the problem of risk
- Exploring legal and scientific notions of causation
- Science in court

This chapter explores some of the connections and differences between law and science in the context of environmental protection. Science is a powerful tool in the search for explanations and understanding of environmental phenomena, and their relationships with human activities. Law, in contrast, serves somewhat different ends: regulation and dispute settlement. It is the frequent overlaps of these two disciplines and their contrasting objectives that are dealt with here. The chapter begins with a general exploration of the nature of science and its congruence or fit with law. There follows a more detailed analysis of three specific areas of interest: the problem of risk, establishing causation, and the role of scientific evidence in court.

The nature of science and its role in law

Science is, broadly speaking, the investigation of objectively verifiable knowledge. It can be defined as a methodology for discovering or approaching truth through theory invalidation.

Since Popper (1934) proposed the criterion of testability (i.e. falsifiability) as the key test for scientific validity, science has proceeded with that criterion in view. Science is not science unless it involves a theory or proposal that could, at least in principle, be shown to be false. More specifically, Popper advanced the hypothetico-deductive view of science: scientific theories are *hypotheses* from which one may deduce statements testable by experimentation and observation. If the statements under investigation are falsified by the observations then the hypothesis is refuted. If not, the hypothesis may (tentatively) be accepted. A key element of Popper's theory is that no scientific theory can, however well supported, be conclusively proved to be correct. There is always the possibility that, in the future, one or more contrary observations may arise.

It is normal for the results of scientific experiments to be subject to mathematical tests for 'statistical significance'. Statistical significance is a device used to rule out chance results. In any investigation there is always a possibility that the result obtained has occurred by chance alone. Tests of statistical significance compare the result obtained with the likelihood of finding such a result by mere chance alone. They seek to identify cases in which there is less than some selected probability that the result obtained is a mere product of chance. Usually scientists opt for a standard of 5 per cent probability of a chance result: in other words, at least a 95 per cent chance that the result obtained is not a product of mere chance. Scientific studies involving small numbers of experimental subjects or observations are particularly likely to fail these tests of statistical significance.

In addition to these tests the norm is for experiments and observations to be subject to peer review followed by journal publication. Peer review allows for the filtering out and rejection of results that are not methodologically sound or which do not advance knowledge. Publication allows for the dissemination of increases of knowledge. Results which have not been subject to tests of statistical significance, and which have not been published in a peer reviewed journal, are not generally considered as part of the body of scientific knowledge.

The above is a fairly conservative and orthodox view of what science is and how it operates. Science in practice differs from science in pure theory. Critics of the orthodox view of science suggest that conclusions are drawn *first* about the subjects and outcomes of scientific enquiry, and that 'science is then used to provide data, theories, and "laws" which support and give legitimacy to these conclusions' (Pepper 1984, 126). We might also observe, with Fisk (1998), that scientific studies are often funded by industries that have a vested interest in uncovering 'evidence' of the safety of their products or services, or avoiding research that might indicate the contrary facts. There is an in-built tendency for science to operate conservatively and endorse the safety of existing substances or products. Even the peer-reviewed publication process is not infallible: journals have an incentive to publish *novel* results which are, by their nature, more likely to be chance correlations, and less incentive to publish repetitions of earlier work which confirm existing theories. Scientists themselves have an incentive to arrive at conclusions that require 'further research', hence further funding. The result is that the scientific literature probably displays an exaggerated degree of chaos and inconsistency.

Some commentators argue for broader and more holistic notions of science in the environmental context. Fisk (1998) decries a view of science as comprising only research that is 'conducted only to the highest standards' and that studies 'that which is known with objective certainty'. This would exclude much of the data that comprises 'environmental studies'. He also warns against affording all statements of technical persons as science. Science, he observes, tells us what is, but does not tell us what to do; this, according to Fisk, requires subsequent risk assessment in the light of the certainties and uncertainties revealed by scientific investigation.

Law, as we saw in Chapter 1, operates on quite different lines to science. Legal 'knowledge' exists in the form of rulings in decided cases or published legislation. Both law and science contain mechanisms for information dissemination but, beyond that, more differences than similarities exist. One key difference is that law does not require or utilise the hypothetico-deductive method. Judges or juries hear evidence put before them by the parties to the case, and the case will be decided there and then by the application of some standard of legal proof: in civil cases usually 'more likely than not' and in criminal cases usually 'beyond all reasonable doubt'. There is no time or space for courts to crystallise theories and direct investigation of statements derived from such theories. There is no requirement for peer review. Legal doctrines or edicts can be changed almost immediately simply through the ruling of a higher court or issuance of amending legislation.

Jackson (1998) has summarised this inherent difference thus:

> A scientist's search for truth is fundamentally at odds with the legal system's approach in resolving disputes. 'The theory of the adversarial system . . . is that if you set two liars to exposing each other, eventually the truth will come out' [citing George Bernard Shaw]. A scientist seeks to discover the truth by uncovering the laws of nature. There will always be more objective scientific 'truth' to be discovered

> on any issues and scientists will never profess to know the ultimate 'truth'. At most, a scientist will tell you what the truth *probably* is, within varying degrees of certainty. A good scientist would rather *not* settle an issue than settle it wrongly, whereas a good lawyer strives to resolve a dispute in the client's favour. While law focuses on settling disputes, science focuses on finding the right answer.
>
> (Jackson 1998, 452)

Jackson's observations may be valid in the context of national adversarial legal systems, but they are less applicable in the context of international environmental law. Several international conventions create forums in which scientific experts can cooperate towards consensus on issues of scientific uncertainty. A good example of this is the Intergovernmental Panel on Climate Change (IPCC) set up by the United Nations Environment Programme (UNEP). This panel has over three thousand scientists who operate a massive peer review process on the global body of literature on climate change, and publish consensus-based reports. In 1994 the IPCC issued a warning that a cut in emissions of around 60 per cent would be needed to stabilise the trend in global warming. Another example, although one which is more politicised and contentious, is the Scientific Committee to the International Convention for the Regulation of Whaling. This body issues scientific assessments of whale stocks and the likely effects of whaling on those stocks. This advice is then utilised as the basis for setting annual catch limits. Unfortunately, the scientists in this context often were unable to agree amongst themselves as to even basic information (e.g. stock levels for any given species) and, furthermore, became politically characterised into pro- or anti-whaling groups.

Wagner (1999) argues that attempts to provide science-based legislative solutions to environmental problems are sometimes misplaced. Science cannot, and does not, have total insight. Reliance on science may, therefore, inculcate an attitude of complacency and over-optimism. Science, she explains, develops incrementally and in a zig-zag manner, producing knowledge and then opening up new gaps in knowledge to be filled. Science has many limits:

- some areas may not have received any or sufficient research to resolve knowledge gaps (e.g. cancer risks from powerlines);
- empirical data are often conflicting and defy a unitary interpretation;
- scientists may have their own 'agendas' and take subjective positions;
- competing theories often exist to explain well researched facts;
- data may merely demonstrate correlation between phenomena, which does not, alone, support findings of cause and effect;

- much science concerning the alleged effects of toxins and carcinogens relies on extrapolation from animal experimentation, which may be unfounded.

Wagner argues that decision-makers lack an appreciation of the limitations of science and, in consequence, overlook the existence of knowledge gaps. This results in unrealistic delegations to agencies – such as requirements to ensure that pollutant levels are 'safe' (e.g. the US Clean Air Act) or that unreasonable risks or adverse effects are avoided (e.g. the US Endangered Species Act). Frangos (1999, 176) similarly notes that, '[r]egulators often pay little attention to what science is capable of determining when they establish criteria for regulatory decision making'. Somewhat ironically it has been observed that fuzziness or a 'veil of uncertainty' in scientific evidence can actually help to *build* the consensus necessary for international legislative action (Haas 1989).

An important distinction can be drawn between science, on the one hand, and law and ethics on the other. Modern science is driven by technological concerns; that is, it has a *know-how* preoccupation (Schumacher 1976; Pepper 1984). Law and ethics, on the other hand, are concerned with prescription and proscription of actions; that is, they have a *know-what* focus. Science cannot tell us what our ethical or legal values should be. Thus a simple analysis of the relationship between law and science might be that society chooses its values, expresses these in laws, and then looks to science to provide the tools to put these legal–ethical values into practice. An environmental example could be the emergence of a consensus that certain species of plants and animals are rare or endangered, and that these species have 'inherent value', followed by the legalisation of this value through new conservation laws, and the instruction of scientists and agencies in actions tending to preserve such plants and animals.

Such a simple unidirectional explanation would, unfortunately, be far from the truth. In reality science, ethics and law are much more tangled and interrelated. More often business directs science, science leads values, and ethics and law trail miserably behind wondering what to do. The areas of genetically modified organisms and human cloning illustrate this complexity. The drive for biotechnology has been orchestrated by a few large multinational corporations, especially Monsanto. Science has achieved the technical results demanded (e.g. crops that are resistant to pesticides). The public have responded differently in the United States (acceptance or support), Europe (suspicion or rejection) and developing countries (acceptance coupled with concern over jobs and the terminator gene). There is, therefore, no global consensus on the ethical or political aspects for international laws to follow. The result is legal conflict and uncertainty.

By predicting the consequences of our actions science provides a tool to help us put our values (whatever they might be) into practice. The kind of question which scientists are asked to tackle, and the interpretation we give to their answers, depends on human value judgments as to what counts as a problem and subjective ideas of risk. Within those limits scientific method strives to describe the behaviour of nature and to predict the effects on nature of our interventions.

Dealing with the problem of risk

Science operates in non-absolute terms. Data is always subject to a degree of error and theories are always susceptible to sudden or incremental change. Scientists, consequently, prefer to operate in terms of *probabilities* and, as a correlative term in the context of harmful events, environmental *risk*. The term 'risk' refers to the probability of occurrence of a harmful event ('hazard') coupled with the magnitude of that event.

In the UK criteria for acceptable and unacceptable risks have been developed by the Health and Safety Executive (HSE) – the agency which deals with risk to workers' health in the workplace. In the document, *The Tolerability of Risk from Nuclear Power Stations* (1988, revised 1992), the HSE propose that an acceptable risk is one which creates one death per million head of population per year in the most vulnerable group of the exposed population. This is equivalent to several common risks such as electrocution, or dying from a fire or gas explosion at home, and is around 100 times greater than the risk of death from a road traffic accident.

At the other end of the scale HSE advise that a risk higher than one death in 10,000 head of population per year is unacceptable. Interestingly, this higher level of risk is not the criterion which the HSE advise planning authorities to use in assessing the acceptability of proposed developments. Instead a figure ten times greater (i.e. one in 100,000), referred to as a 'substantial risk', is recommended for assessing unacceptable risks. Setting the 'applied' criterion at an order of magnitude higher than the 'real' acceptable risk allows for uncertainty in the results of risk assessment and possible future changes in the use of the plant.

Where the magnitude or type of harm posed by the probabilistic event is low, much higher risks of occurrence may be acceptable, or be required to justify prohibition of development. Potential economic impacts may be less serious than potential loss of life. *Envirocor* v. *Secretary of State for the Environment and British Cocoa Mills (Hull) Ltd.* [1996] provides a good example of the difficulties of choosing the appropriate numeric risk level in cases not involving a threat to life. In this case a Planning Inspector refused an appeal against a decision by Humberside County Council rejecting an application for planning permission to operate a waste transfer station on a site next to British Cocoa Mills' factory. The Inspector found that, in the event of an accidental spillage of chemicals on the proposed waste transfer station site, the risk of tainting of BCM's food products from malodorous emissions would be too great. The Inspector concluded that a risk in the order of one in 300,000 of a 'worst case scenario' (i.e. an undetected tainting which might threaten the very future of the factory) could not be ignored and that such a risk justified the Council's original decision not to grant planning permission. Envirocor appealed successfully. The judge ruled that the Inspector's approach had been unlawful. The BCM plant and the waste site would be likely to coexist for around thirty years: a finding that a 'worst case' tainting incident was 'likely to occur' would presuppose a risk of around one in 30 – a risk much higher than that suggested by either side in the litigation.

This decision is open to criticism: the judge seems to have interpreted the notion of 'likely to occur' as 'virtually certain to occur' (i.e. nearly a 100 per cent probability of occurrence). Tromans (1995) comments that the *Envirocor* decision illustrates the difficulty for courts faced with risk-assessment decisions when there is no input from a body with environmental risk assessment expertise, such as the Environment Agency or the HSE.

The attempt to create precise mathematical calculations of environmental risks can be difficult, if not impossible, in certain contexts. Environmental systems contain too many parameters, too many imponderables, too many possible causal links for anything like absolute certainty in calculation of risks. Some critics go further and dispute the very basis of risk assessment, especially of long-term or low-dose exposures (Frangos 1999; Shere 1995)

The UK's Environment Agency has developed a 'tiered' and 'iterative' risk assessment strategy. The process is tiered because it involves identification of key risks and priorities. Only if a decision cannot be made based on that information is a more detailed approach taken. The process progresses from key risks to more detailed considerations and requires repeated re-examination of assumptions (Environment Agency 1999, 10). The process, which is similar to that applied by the US National Research Council (Kubasek and Silverman 2000, 213), typically involves the following stages:

1 Hazard identification (i.e. what hazards are present and what are their priorities?).
2 Exposure assessment (i.e. how might the receptors become exposed to the hazards and what is the probability and scale of the exposure?).
3 Risk estimation (i.e. given probability and scale of exposure, what is the probability and scale of harm?).
4 Risk characterisation (i.e. how significant and/or acceptable is the risk and what are the uncertainties?).

Of these the fourth stage is the least objective and, given judicial deference to agency judgments, the least susceptible to judicial challenge.

Because of the lack of public involvement in agency decision-making processes, and lack of transparency of such processes, it can often be difficult for the public to be certain that full consideration has been given to the environmental risks inherent in a given activity. This information deficit can prompt legal challenges. In *R.* v. *Environment Agency, ex parte Turnbull* (2000) for instance, the applicant, Turnbull, challenged the decision of the UK's Environment Agency granting approval for an incinerator for the carcasses of cattle affected by Bovine Spongiform Encephalopathy (BSE). The challenge failed because it became apparent, in court, that detailed behind-the-scenes consideration of the risks involved had already taken place. One could argue that environmental legislation which provides for risk assessment should place agencies under an obligation to publish summaries of the risk assessment process and outcome that has taken place, thereby lessening the perceived need to challenge such bodies in the courts.

One question which often surfaces in the environmental risk debate is *whose* assessment of risk should count as legitimate – that of the public, that of scientists, or both? It has often been observed that public risk perceptions may not match those made by experts or scientists. Public risk judgments, unlike scientific risk assessment, tend to take social, political, cultural and ethical factors into account (Frangos 1999). Public assessments of risk are characterised by 'rule of thumb' or binary type decisions, greater reaction to more 'prominent' issues that stand out from their background, greater concern for things which occur 'close to home', distrust of experts, and non-mathematical modes of reasoning (Breyer and Heyvaert 2000, 291). The public also tend to give priority to the avoidance of large singular catastrophic events, or irreversible damage, even if these, from a strictly mathematical point of view, are less risky than multiple small damage occurrences. The classic example of this is the fear of death by flying (objectively low, subjectively high) compared to the fear of death from vehicle accident (objectively high, subjectively low). The public tends to rate harm from nuclear power as more risky than harm from smoking, even though most people know that more deaths are caused by smoking (Frangos 1999, 177).

A New Zealand case involving radio wave transmissions illustrates this divergence and its resolution in a legal context. In *Shirley Primary School* v. *Telecom Mobile Communications Ltd* [1999] the applicants objected to the construction and operation of a cellular radio base station on land adjacent to their school. School officials were not convinced by the telecom company's assurances that the risk to the children's health was minimal. The judge, however, refused to overturn the decision granting permission for the transmitter and, in so doing, indicated a lack of confidence in public perceptions of risk:

> As a link between the adverse (physical) health effects as we have found them, and the psychological effects . . . we observe that there is often a large gap between scientists['] and the public's assessment of risk. Scientists attempt to calculate risk on a probabilistic basis, whereas the public is swayed by other factors or, possibly, by the same factors viewed in a different way. One aspect of this is that most people have considerable difficulty understanding the mathematical probabilities involved in assessing risk. . . People consistently overestimate small probabilities. What is the likelihood of death by botulism? (One in two million). They underestimate large ones. What is the likelihood of death by diabetes? (One in fifty thousand). People cannot detect inconsistencies in their own risk-related choices.

The judge in the *Shirley* case went on to support the company's assessment that the risk posed by radio frequency radiation was low and acceptable.

Should public perceptions of risk count if they are not at one with scientific views or are, in some sense, 'irrational'? This issue arose quite early in America in a sequence of cases concerning the risk of infection from hospitals established in populated areas. In *Everett* v. *Paschall*

(1910), for instance, the court upheld an injunction preventing the operation of a tuberculosis sanitorium in a residential area on the grounds that where there exists a 'public dread',

> [t]he question is, not whether the fear is founded in science, but whether it exists; not whether it is imaginary, but whether it is real, in that it affects the movements and conduct of men.

On this interpretation public perception of risk is counted as a harm in itself which should not have to be endured. Of course, this position has not endured – it is no longer possible to mount a legal challenge to a development simply on the grounds that a real if misplaced fear of harm will occur. In the UK case *Gateshead Metropolitan Borough Council* v. *Secretary of State for the Environment and another* [1995] the judge was asked, in effect, to consider whether a planning consent for an incinerator could lawfully be turned down because of public fears about inadequate regulation of such a plant. He commented:

> Public concern is, of course, and must be recognised by the Secretary of State to be, a material consideration for him to take into account. But if in the end that public concern is not justified, it cannot be conclusive. If it were, no industrial development – indeed very little development of any kind – would ever be permitted.

To this rule several exceptions exist. Where a negligent act induces a fear of a risk of harm that, in turn, precipitates a recognisable psychiatric condition, then liability will arise for that harm. In *Andrews and others* v. *Secretary of State for Health*; *Re CJD Litigation* (1998) victims who had suffered depression and nervous breakdowns as a result of receiving growth hormone treatment carrying a risk of transmission of Creutzfeldt-Jakob disease (CJD) succeeded in claims for damages. Once a nuisance has been objectively established, courts may allow for recovery of 'emotional anguish' or stress arising from foreseeable escapes of pollutants (*Ayers* v. *Township of Jackson* (1983)). There has also, recently, been successful litigation in the US for the reduction in property value resulting from the fear, rational or otherwise, of an increase in cancer risk as a consequence of electromagnetic fields from high voltage power lines (Young 1991).

In contrast to their attitude to public assessments of risk, courts often show a willingness to support the risk assessments made by statutorily mandated agencies, even where the science and data that they employ are unclear. In *Ethyl Corporation* v. *EPA* (1976), for instance, the petitioners challenged a ruling by the US EPA regulating the introduction of lead additives to vehicle fuel. Although lead is well known to be toxic it was hard for the EPA to establish precisely what proportion of lead exposure for the affected population resulted from the insertion of lead in fuel. There was, as a consequence, considerable uncertainty as to the health benefits that would accrue from the regulation. The court, upholding the EPA's action, observed that regulators were not blessed with 'a prescience that removes all doubt from their decision making', and that, in the context of precautionary legislation, regulators should not be expected to achieve certainty or rigorous proof of cause and effect before taking action. This decision has been followed in numerous lower court decisions and upheld by the Supreme Court (e.g. *Industrial Union Department, AFL-CIO* v. *American Petroleum Institute* (1980)) and can be seen as an early manifestation of the precautionary principle (see Chapter 4).

American courts not only tend to support agency risk assessments, but also to positively require regulation when environmental risks exist and supporting legislation mandates the amelioration of such risks. In *NRDC* v. *United States EPA*, for instance, the EPA had refused to set standards for vinyl chloride on the grounds that only a zero-emission standard made sense. The court rejected this paralysis: the Clean Air Act required the EPA to regulate hazardous air pollutants 'at a level that provides an ample margin of safety to protect public health'. By refusing to set a standard the EPA had committed the error of substituting technological feasibility for health. The court ordered standards to be set based on expert judgment regarding the emissions that would result in an 'acceptable' risk to health.

Environmental risks may be more or less acceptable depending upon whether they are voluntarily accepted or imposed. Voluntary risks are part of an individual's assessment of the trade-off between risks and benefits. As such they are acceptable to the individual assuming the risk. Douglas and Wildasky note that

> [a]lthough his eventual tradeoff may not be consciously or analytically determined, or based on objective knowledge, it nevertheless is likely to represent, for that individual, a crude optimization appropriate to his value system.
>
> (cited in Silver 1986, 67)

Obviously, individuals can only make judgments about environmental risks if they are provided with clear and pertinent information about the same. This is one reason why laws giving rights of access to environmental information are so important. European States have an obligation under Article 8 of the European Convention of Human Rights (right to private and family life) to provide access to information about risks of hazardous activities to those engaged in those activities (*McGinley and Egan* v. *UK* (1998)). Where information about a serious risk is not provided then Article 8 may be infringed. In *Guerra and others* (Thornton and Tromans 1999) the European Court of Human Rights found infringement of Article 8 where the Italian authorities had failed to provide local people with information about the risks from a chemical factory that had already suffered serious accidental releases.

Involuntary risks are not, generally, as acceptable to individuals. One of the problems of involuntary risks is that they interfere with, or are imposed on top of, a person's individualised risk–benefit assessment. They are also an infringement of autonomy, forced on individuals without the possibility of acceptance or rejection. Involuntary risks are often 'public' risks; that is, 'threats to human health or safety that are centrally or mass-produced, broadly distributed, and largely outside the individual risk-bearer's direct understanding and control' (Huber 1985, 277). The division between voluntary and involuntary risks, although conceptually clear, is often difficult to make. Is living in an area of high pollution a voluntary or an involuntary risk, given the difficulties and impracticalities (especially for children) of moving to a different place?

However, this does not mean that all involuntary risks are unacceptable. Involuntary risks may be justified by vastly outweighing benefits, or if they are low in comparison to background risks or voluntary risks of the same or similar kind. For example, most radiation risk is either involuntary background risk (i.e. naturally occurring sources of radiation) or voluntary risk (voluntarily accepted medical interventions – for example, x-rays and scans). Courts are sometimes sensitive to this categorisation of risk. In the *Shirley* case (see p. 203) the court observed:

> We have to recognise how much EMR citizens of New Zealand are exposed to both voluntarily and involuntarily . . . everyone in the whole world is exposed to EMR all the time. That includes exposure to the most dangerous EMR which is high-frequency ionising radiation (such as cosmic rays). At lower frequencies there is ultraviolet light and then the narrow band of visible light with frequencies of between 10^{14} and 10^{15} Hertz. The important and conspicuous EMR we all receive is direct from the sun. Sunlight gives each and every living thing a continuous exposure of about 80,000

> [micro] W/cm^2. Below the frequencies of visible light there is no danger from ionising radiation. This radiation can of course still be dangerous – it contains enough energy to cause hearing or thermal effects. However, greater exposures are needed at lower frequencies to cause those effects. So there is nearly nothing special about radio frequency (RF) radiation – it is just one of the many forms of EMR that humans have evolved to live with. However, the background natural level of RFR is very low. It is only in the last 100 years that we have become exposed to much more 'unnatural' ie human-generated RFR. Now we receive it from televisions, microwave ovens, electric blankets, visual display units and of course cellphones.

Another characteristic of risk which the courts take account of is temporal immediacy. A risk of something occurring in a hundred years may be thought of as less objectionable than a risk of the same event occurring tomorrow, even if it unclear why, on ethical grounds, this should be so. The common law is generally unwilling to involve itself in protection of individuals against remote or future risks. This reluctance is reflected in the limitation of availability of prospective ('*quia timet*') injunctions to cases of *imminent* and *irreparable* harm. In *Salvin* v. *North Brancepeth Coal Co.* [1874], for example, the plaintiff sought an injunction to prevent a nearby colliery from beginning to operate its coal ovens. The court rejected the claim: at the time of application to the court there was no 'visible' or 'sensible' harm to the plaintiff's crops as required by law. The courts also often show faith, perhaps even a misplaced faith, in the ability of science and human ingenuity to produce solutions before apparent risks materialise. In *Attorney-General* v. *Mayor, Aldermen and Corporation of Kingston-upon-Thames* (1865) the conservators of the River Thames challenged a proposal to discharge Kingston's untreated sewage into the river. The court heard that the sewage might, at some time hence, pollute the town's water supply or interfere with fishing. The court, refusing to grant an injunction, observed:

> Something which might happen in a hundred years' time would not justify the interposition of the court; inasmuch as by that time a variety of chemical continuances might be found to prevent the evil, and a variety of contrivances might be discovered by which the very thing that now was being done, might be done without the evil effect anticipated.

Many environmental risks show the characteristic of latency; that is, that the harmful consequences do not materialise for a considerable period of time after the originating event. Pollution of groundwater, for instance, often takes many years to appear after a long period of downward

discipline. Furthermore, as other writers have noted (e.g. Diamond 1968), all expert witnesses have some set of personal values which consciously or unconsciously colour their evidence. Several studies have shown that, despite the rigour of their research and observations, scientists often do not make good witnesses and can suffer from focus by legal advocates on apparent weaknesses in science as a form of knowledge (Yearly 1991, 140). These can include 'uncovering' apparently rigorous scientific techniques as 'mere conventions' and demonstration that scientific models used fail to take account of real-world variables.

One alternative to the partisan use of scientific expert witnesses is for the court to appoint independent experts, either as witnesses, or as technical advisors (Jackson 1998). The advantage of an independent expert is that he or she is not concerned with defence of his or her own reputation and point of view, and is not embroiled in the project, implicit in an adversarial legal system, of rubbishing the opposition. Of course, the use of independent experts is itself problematic. Who is to determine which expert shall be appointed. Can one be confident, given what has already been said, that any such expert would be neutral between the parties?

Key points

- The processes of scientific inquiry have been criticised as not being conducive to effective environmental protection.
- The notion of risk is problematic. Legislation and courts are cognisant of the difference in risk perception between public and 'experts', but usually favour the latter over the former.
- Establishing causation is a major hurdle for those wishing to bring an action for damages resulting from some environmentally induced illness or injury. Different rules of proof apply in different states but, in general, courts have made sympathetic efforts to assist claimants overcome the impossibility of full scientific proof.

Further reading

Bosselman, F.P. and Tarlock, A.D. (1994) 'The Influence of Ecological Science in American Law: An Introduction', *Chicago-Kent Law Review*, vol. 69, pp. 847–73, and Frangos, J. (1999) 'Environmental Science and the Law', *Environmental and Planning Law Journal*, vol. 16, no. 2, pp. 175–81, both provide good general discussions of the interface between environmental law and environmental science.

Cranor, C.F., Fischer, J.G. and Eastmond, D.A. (1996) 'Judicial Boundary Drawing and the Need for Context-Sensitive Science in Toxic Torts after Daubert v. Merrell Dow Pharmaceuticals, Inc., *Virginian Environmental Law Journal*, vol. 16, pt 1, pp. 1–77. This explores the role of the judge as the gatekeeper of good science in environmental claims cases.

Discussion questions

1. Are law and science *really* so different?
2. How much weight should be given to public perceptions of environmental risk, and how should such perceptions be factored into the overall political and decision-making process?
3. Is it fair that victims of pollution injuries have to establish proof of causation? Do existing legal rules strike a fair balance between protection of the industrial operator and protection of the individual in this matter?
4. Can we rely on scientific expert witnesses to discover the 'truth' in environmental litigation?

8 Environmental ethics in law

- Schools of environmental ethics
- Anthropocentric ethics
- Biocentric ethics
- Ecocentric ethics
- Ethics in hunting, conservation and pollution
- The place of ethics in environmental law

This chapter explores the relationship between environmental ethics and law: two distinct yet potentially related means of achieving the shifts in human behaviour that are necessary to prevent or slow the trend of significant environmental damage. The discussion examines, specifically, the connection between law and ethics in the areas of international law, conservation, hunting, and pollution control.

A taxonomy of ethics

Environmental ethics are systematic comprehensive accounts of the moral relations between human beings and the environment (Des Jardins 1997). They are *normative,* i.e. they tell us what is 'right' (appropriate actions) or 'good' (appropriate outcomes). As such they provide a target for human endeavour, and a bench-mark against which other social systems, such as law, can be measured.

Environmental ethics have, in the main, evolved as specific applications or extensions of 'human' ethics, making it necessary to have some familiarity with the latter in order to make sense of the former.

At the risk of some oversimplification, we can divide human ethics into four main groups: consequentialist ethics, duty-based ethics, rights-based ethics, and virtue ethics. *Consequentialist* ethics identify outcomes or goals (consequences) which rational agents ought to try to attain. Acts are 'right' if they achieve these 'good' outcomes.

The best-known consequentialist ethic is *utilitarianism*. In Bentham's version this calls for the maximisation of pleasure and the minimisation of pain:

> Nature has placed mankind under the governance of two sovereign masters, *pleasure* and *pain*. It is for them alone to point out what we ought to do . . .
>
> (Bentham 1948, orig. 1789, 1)

> An action then may be said to be conformable to the principle of utility, or, for shortness sake, to utility, (meaning with respect to the community at large) when the tendency it has to augment the happiness of the community is greater than any it has to diminish it.
>
> (Bentham 1948, orig. 1789, 3)

For Bentham all pleasures count equally. But for his disciple, John Stuart Mill, civilised pleasures are worth more than base pleasures: 'better to be a human being dissatisfied than a pig satisfied: better to be Socrates dissatisfied than a fool satisfied' (1972, orig. 1863, 10). Mill was influenced by Whewell's criticism that, were it not so, it would be our duty to populate the earth with a surplus of happy cats, dogs and pigs at the expense of human pleasure.

Mill's utilitarianism is also notable for its suggestion that happiness maximisation requires not attendance to the principle itself, but rather obedience to subordinate rules. This derives, in part, from Mill's recognition (and fear) that happiness might, on occasions, be greatly increased by apparently immoral acts such as assassination.

In modern times changes to utilitarianism include replacement of the simple notion of 'pleasure' with concepts such as 'interests', 'preferences', 'welfare', or a whole basket of 'goods' (Moore 1903) such as knowledge and beauty. Modern utilitarianism finds particular expression in the field of welfare economics, which investigates the circumstances and preconditions necessary for the maximisation of social welfare.

Some important objections to utilitarianism (generally, see Smart and Williams 1973) can be enumerated:

- Predictions of pain or pleasure are difficult to make and frequently incorrect. Only a few of the more direct consequences of any act can be predicted with any confidence.

- It is not clear that there can be any objective verifiable measurement of pleasure or pain that could be used to make the interpersonal comparisons necessary for deciding between proposed courses of actions.
- Utilitarianism can lead to repugnant conclusions. In a community of sadists, public child torture would be morally laudable, indeed obligatory. This is because utilitarianism treats humans merely as 'pleasure/pain vessels'; that is, it contains no grounds on which to respect human life *per se*.
- Utilitarianism which calls for comparison of higher and lower pleasures cannot provide any rational criteria for this distinction (other than intuitions) and runs the risk of 'ethical snobbery' in which the pleasures of the elite are arbitrarily preferred to the pleasures of the masses (Mill 1972, orig. 1863, xvi).
- Utilitarianism has difficulty accounting for rights and duties that arise from past actions, and from special relations between people (e.g. promises, punishment and retribution, marriage and friendship).

Duty-based ethics construe 'rightness' not in terms of consequences or outcomes, but in terms of *duties* – such duties being of a self-evident or self-substantiating kind. The best-known duty-based ethic is that developed by Immanuel Kant (Paton 1948; O'Neill 1991). Kant held that people are free agents capable of making decisions, setting goals and acting according to reason (hence morality). A person acts ethically when he or she acts out of a realisation of moral duty. A person who acts from a sense of moral duty treats others with respect, respecting their autonomy.

Kant's ethics can be thought of as encapsulated in two versions of his 'categorical imperative':

> Act only on the maxim through which you can at the same time will it to be a universal law.
>
> Act so that you treat humanity, whether in your own person or in that of another, always as an end and never as a means only.

The first formulation rules out any ethical prescriptions which are internally contradictory – for example, (to use Kant's example), a maxim of making false promises. The latter formulation requires us to abstain from treating persons instrumentally (i.e. it rules out acts which are disrespectful of persons).

Criticisms of Kant's ethic (see O'Neill 1991, 181) include observations that:

- The categorical imperative can be seen as an empty shell: quite trivial acts could be universally willed and performed without treating others as means. But would they be 'right' or would the outcomes be 'good'?
- The ethical principles derived from Kant's ethics are too abstract to act as a guide to action.
- Kant's ethics may lead to rigidly insensitive rules which cannot take account of difference between cases.
- The ethic may collapse into consequentialist ethics since regard to consequences may be necessary in giving content to the categorical imperative.
- Maximising pleasure may simply be *better* than respecting every individual's autonomy or acting out of duty. (Would it really have been wrong to assassinate Hitler? Wouldn't it be best to use people as a means if that left society vastly better off?)
- 'duty for duty's sake is absurd' (Mackie 1984, 171).

Rights-based ethics have strong links with notions of liberty and freedom. The concept of 'rights' is, itself, contested academic territory. Rights can be thought of as freedom from interference. Thus, Hohfeld (1919) states that a *right* (in his terminology a 'claim right') requires a correlative *duty* on the part of others not to prevent the exercise of that right. For instance, a right to drive my car down a public road implies a duty on others not to stop me, either directly or indirectly, by blocking the highway. J.L. Mackie (1984) argues that rights are the very thing which ethical conduct ought to promote and protect, and that (as with Mill) the central right in a right-based ethic ought to be the 'right of persons progressively to choose how they shall live'. The justification for such freedoms is often that they are preconditions for the human good. Finnis (1980), for example, considers rights, especially human rights, to be valuable promoters of human flourishing and, as such, valuable as part of the vocabulary of 'practical reasonableness'. Gewirth (1981) has it that rights (to certain things and to be free from certain interferences) are necessary for persons to function as moral agents, to exercise autonomy and make rational choices.

It is possible that the state is the body whose interference must most significantly be resisted. In his essay, *On Liberty*, J.S. Mill argues (but not in such terms) that individuals should be free from state interference, except where this is necessary to protect the interests of others from harm. Dworkin characterises rights as 'trumps over some background justification for political decisions that state a goal for the community as a whole' (1984, 153). For Dworkin rights grant priority to individuals *over* claims to social welfare.

It is important to be clear that possession of a moral right is not the same thing as possession of a legal right. As we have seen, a moral right consists of a claim to a certain freedom reflected in some correlative duty, whereas a legal right (Stone 1972) has four components:

1 there must be a public authoritative body that is prepared to review actions colourably inconsistent with the 'right';
2 the thing or person that holds the right must be able to institute legal actions at its own behest;
3 in granting relief the court must take account of injury to the right holder;
4 the relief granted by the court must run to the benefit of the right holder (not some other person or body).

Rights-based ethics do not necessarily imply equality. Rawls (1972), in a social contract style of reasoning, maintains that justice comprises those principles for social institutions that would be agreed to by persons acting behind a 'veil of ignorance'; that is:

- each person is to have an equal right to the most extensive basic liberty compatible with a similar liberty for others, and
- social and economic inequalities are to be arranged so that they are both (a) reasonably expected to be to everyone's advantage and (b) attached to offices and positions open to all.

As we can see Rawls is explicit in his denial of any right to social and economic equality (so long as conditions (a) and (b), set out above, are satisfied). This theme is shared by Nozick's (1974) theory of the 'minimal state', which includes an assertion of a general right to appropriate and hold property that one has produced or justly acquired. A resource is justly acquired if 'the situation of others is not worsened' (Nozick 1974, 175). Redistribution on welfare grounds infringes that right, unless the original circumstances of acquisition are themselves unjust.

Arguments against rights-based ethics include the views that:

- Rights are nonsense and natural rights are 'nonsense upon stilts' (Bentham 1970).
- All attempts to give good reasons for the existence of rights fail, or collapse into other arguments such as utility (see, e.g., MacIntyre 1985, 69).
- Rights imply an individualistic rather than a communal conception of humanity (MacIntyre 1985).
- Rights emphasise the male value of individual autonomy and de-emphasise the female value of caring (Gilligan 1982).

The fourth school of ethics, *virtue ethics*, seeks to answer questions such as 'what is the life worth living?' or 'what kind of person should I be?' A virtuous person will possess certain characteristics (e.g. courage, honesty). This kind of ethic does not necessarily generate precise formulae for action (e.g. maximise pleasure), although it is capable of doing so. For example, Aristotle, the 'father' of virtue ethics, suggested that a good person would live life according to the 'the mean': for example, be neither too bold nor too timid, neither too kind nor too hard. A virtuous person might be one who generally treats others with respect, or acts out of compassion. Virtue theory is, in part, a response to a perceived weakness of standard ethics (utilitarianism, duty-based ethics and rights-based ethics), which is that 'it is theoretically possible that a person could, robot-like, obey every moral rule and lead the perfectly moral life' (Pence 1991). Such a life could be empty, isolated and unconnected. Some virtue theorists (e.g. MacIntyre) have argued that virtues can only flourish when a person belongs to a society or group in which his or her character is linked to the history or moral traditions and practices of that group.

Criticisms of virtue ethics occur both from within the other three main schools (e.g., for a utilitarian, 'what is the use of courage and honour if it does not promote happiness?') or internally (e.g., virtue theory has no master principle and so can lead to adulation of morally repugnant characters).

Environmental ethics

As we shall see environmental ethics draw strongly on the fundaments of human ethics; that is, notions of utility, good, duty, respect, rights and virtue. But for the purposes of our discussion it is best to consider them as classified according to the subject matter of their concern. We can, adopting this taxonomy, distinguish between ethical theories concerning (a) proper treatment of the environment for human ends ('*anthropocentric ethics*'), (b) proper treatment of animals and plants ('*biocentric ethics*'), and (c) proper treatment of species and ecosystems ('*ecocentric ethics*').

Anthropocentric ethics

According to John Passmore (1980) sufficient reasons can be given for protecting the environment by applying ordinary human ethics and pragmatic reasoning about the human good. Indeed, this is the kind of 'ethic' which is commonly encountered in environmental policy literature; that is, that halting phenomena such as climate change, ozone depletion and species extinction is required of us because of our ethical relations one to another. Enlightened thinkers will take into account long-term human benefits of preserving biodiversity (Wilson 1992) and the pleasures to be derived from nature contemplation (Mill 1891).

A common manifestation of utilitarian anthropocentric ethics, although one which usually keeps its provenance hidden, is *environmental economics*. A central tenet of this discipline is that society should evaluate decisions about protection or destruction of the environment by reference to the notion of *human welfare*. Actions that maximise human welfare are described as *efficient*. Efficiency is generally taken to require laws that do not totally proscribe pollution but, rather, moderate pollution to an optimal amount (Baxter 1974; Pearce and Turner 1990). Conservation, similarly, is only deemed necessary in so far as the objects conserved contribute to human satisfactions:

> My criteria are oriented to people, not penguins . . . Penguins are important because people enjoy seeing them walk about on rocks . . . I have no interest in preserving penguins for their own sake.
>
> (Baxter 1974, 5)

In an effort to attain efficient solutions environmental economists propose techniques such as cost–benefit analysis and the monetisation of environmental harm through processes such as Contingent Valuation (CV) methods – for example, as 'willingness to pay' for conservation or 'willingness to be compensated' for environmental harm. However, not only are CV methods methodologically suspect (e.g. Note 1992; Cichetti and Peck 1989) but, as Sagoff (1988) points out, many people find such processes immoral and ethically inappropriate.

Utilitarianism and its 'daughter' welfare economics are not the only positions that are encountered in anthropocentric ethics. Indeed, quite radical positions can be attained without needing to assert the ethical considerability of individual animals, species or ecosytems. One can, for example, take the view that to promote beauty is good, that nature is beautiful, and that therefore nature requires our protection (Hargrove 1989). One can adopt the virtue-ethics position that 'failure to appreciate the natural scene is as serious a human weakness as a failure to appreciate works of art' (Passmore 1985, 217). This may lead to the view that wilderness should be retained as an arena for human interaction, through outdoor pursuits such as hiking and camping, from which rounded and fully developed human beings can develop (Sax 1970). A further reason commonly given for preserving nature is that people value it (not, note, that it is *in itself* valuable).

A problem with anthropocentric environmental ethics is that the basic premise – what is good for humans is good for nature – may not be true, or may only be true in certain cases. Perhaps long-term preservation of life on earth requires the human species to attain massive reductions in its population. Perhaps those aspects of nature which bring the most human pleasure, and which are therefore often taken to be worthy of conservation, are in fact the *least* important when the environment's own interests are taken into account (soil bacteria may be ecologically more important than tigers or penguins).

A further, and perhaps more damaging, criticism is that ethics which take nature to be valuable as a means to fulfilling human needs, desires or obligations, miss the point completely. We would not find it acceptable to consider ethical duties to or rights of children as a mechanism for fulfilling moral relations between adults. Why then should the environment and its components be left in the same lowly position? Ultimately we may be left with the feeling that conventional intra-human ethics cannot account properly for the sense that nature is worth protecting – not only for human ends but for itself.

Biocentric ethics

Biocentric ethics answer the criticism, given immediately above, by asserting that animals (and in some cases plants) have *moral considerability*; that is, that they count as ethical subjects in their own right, rather than as an offshoot of human concerns.

Ethics that deal with the position of animals are often based on the notion of *sentience* (i.e. the capacity to experience pleasure and pain). Warnock (1971) and Frankena (1979) agree that sentient creatures can, because of this very capacity, be treated rightly or wrongly. Ethics focused on sentient animals include duty-based ethics (McCloskey 1979; Midgley 1983), Regan's rights-based approach (1983) and Singer's utilitarian perspective (1976). Of these Regan's and Singer's theories are the most developed.

Regan argues for animal rights by asserting (a) that human ethics are properly based on rights and (b) that, as higher animals are not different from humans in any morally relevant way, they too have rights. Humans have rights, including a right to life, because they have 'inherent value', which in turn flows from the fact that they are 'subjects of a life' (i.e. possessed of beliefs, desires, perception, memory and other mental faculties). Higher animals (specifically mentally normal mammals of a year or more in age) also possess all these qualities. Hence higher animals must also have (at a minimum) a right to life.

The animal-rights position is controversial. In addition to the general objections to rights-based ethics canvassed above, some deny that it makes any sense at all to speak of animals possessing rights, as rights are by their very nature limited to beings with linguistic abilities (Francis and Norman 1978; Frey 1980) or sentient beings (McCloskey 1979). Regan can also be criticised for failing to provide a defensible solution to the problem of what to do when animal rights and human rights conflict. Regan's solution (that a human right to life should prevail over animals' rights to life) seems inconsistent with his assertion that both humans and animals are possessors of inherent value (Gruen 1991). We might also point out that, whilst not a criticism of Regan's theory itself, animal rights based on the qualities of only sentient animals cannot form the basis of an ethic which would show us how to relate to the greater proportion of the animal kingdom.

Utilitarian animal ethics can be traced back to Jeremy Bentham who predicted that

> The day may come when the rest of animal creation may acquire those rights which never could have been withheld from them but by the hand of tyranny . . . a full-grown horse or dog is beyond comparison a more rational as well as more conversable than an infant of one day, or a week or even a month old. But suppose the case were otherwise, what would it avail? The question is not, can they reason? nor, can they talk? but can they suffer? Why should the law refuse protection to any sensitive being? The time will come when humanity will extend its mantle over everything that breathes.
>
> (Bentham 1948, orig. 1789, ch. 17, para. 4)

Utilitarian animal ethics have been advanced by Peter Singer (1976) who argues that the exclusion of animals from moral consideration is 'speciesism': arbitrary discrimination on a par with discrimination on grounds of race or gender. According to Singer a fundamental principle of moral theory is that all moral interests should be given equal consideration. All creatures with capacity to suffer have a moral interest – at least an interest in not suffering. Singer's conclusion is that (at least) serious animal suffering should not be allowed merely for human convenience.

Critics of animal utilitarianism have at their disposal all of the criticisms which pertain to utilitarianism in general, and a few more besides. For instance, there is the old issue, first raised by Descartes, of whether animals can feel pain and/or pleasure. If so which animals, and is pain and pleasure in this context qualitatively or quantitatively the same as human pain? As with Regan's rights model, difficulties also arise when we seek to take animals into account in the 'hedonic calculus': we might, for example, be forced to conclude that it is better to save the life of a happy dog than a miserable human (Gruen 1991, 348). Fervent animal supporters may also object to the result, implied by utilitarianism, that the painless killing of animals is not unethical if it increases net welfare or happiness. This could imply sanctioning of animal experimentation, xenotransplantation. Indeed, if human satisfaction from killing and eating animals is great enough then factory farming must be allowed to continue (Regan 1983). The utilitarian approach is also vulnerable to the 'replaceability argument': there seems to be no utilitarian objection to killing creatures painlessly if we replace them with equally happy substitutes (Pluhar 1995, 185).

Animal-rights and animal-utilitarianism are both subject to the objection that they fail to account for the moral status of 'lower animals'. An ethic which manages to incorporate a wide range of animals must either postulate moral equivalence or some form of moral hierarchy for animals possessing widely differing phylogenetic characteristics. Sumner (1981) considers that ethical priority depends upon sentient consciousness, and that rationality enhances sentience, resulting in a hierarchy in which moral standing is afforded in varying degrees. Humans, having the most developed sentience, count the highest. Mammals, especially primates, count for a lot. Animals such as reptiles and birds are significant but rank lower. Insects and micro-organisms count for virtually nothing. According to VanDeVeer (1979) all organisms possess equal moral *considerability* (Goodpaster 1978), but not equal moral *weight*; the moral weight to be afforded to any given organism being a function of (a) the type of interest that is at stake and (b) the organism's psychological capacities. Ethical hierarchies such as these accord with most people's ethical intuitions.

Several philosophers have advanced the view that plants matter ethically (e.g. Arbor 1986 and Stone 1972). Some go further and claim that all living organisms – plants and animals – are important moral subjects (e.g. Schweitzer 1923 and Taylor 1986). In Taylor's case this is because all plants and animals have 'inherent worth'. The criterion for a thing possessing inherent worth is that it has 'a good of its own'; that is, one can conceive of actions being either bad or good for that thing. Cutting down a tree, whilst causing no pain and frustrating no desires or preferences, nevertheless interferes with that tree's 'good'.

According to Taylor recognition of the inherent worth in every living organism leads naturally to the general attitude of *respect for nature*. The person who respects nature will accept a number of duties:

(a) the duty of 'non-maleficence' (a duty not to do harm to any entity in the natural environment which has a good of its own);
(b) the duty of non-interference (a 'hands-off' policy in relation to biotic communities);
(c) the duty of fidelity (a duty not to deceive – for example, by trapping, or to break trust that a wild animal has had to place in us in a situation of our making);
(d) the principle of 'restitutive justice' (a duty to compensate a moral subject where the subject has been wronged by the moral agent).

Taylor's biocentric ethic is duty-based, but it is possible to argue for a broad and inclusive biocentric ethic from any number of foundations (e.g. rights-based and virtue-based positions). For instance, in his famous essay, 'Should Trees Have Standing?', Christopher Stone (1972) called for recognition of the legal (not necessarily moral) rights of trees, valleys and rivers. Stone thought that it is possible to identify violations of the rights of a valley or a tree (e.g. massive development of a pristine valley, or felling of a tree, respectively), and that these entities could, through representative humans, apply to a court for judicial review of such infringements. Criticisms of this position (e.g. Huffman 1992) include doubts that we can say what is in the interests of a tree or valley:

> If we could somehow persuade ourselves that it is not silly to ask whether a 600 year old Douglas fir tree would prefer to provide the supports for a revered structure like Timberline Lodge on Oregon's Mount Hood or to crash to the ground in a windstorm and slowly decay into nothingness, we can simply never know the answer.
>
> (Huffman 1992, 59)

This point, however, probably confuses the notion of preference (only applicable to higher animals) with that of inherent worth (applicable to all organisms).

Ecocentric ethics

A defect or gap in both anthropocentric and biocentric environmental ethics is their failure to provide any ethical grounds, other than indirect considerations, for conduct towards 'wholes' such as communities, populations, species, and ecosystems. Various ethics, which can loosely be grouped together as 'ecocentric', fill this gap.

Several philosophers have proposed the moral considerability of species. Holmes Rolston III maintains that an endangered species counts ethically because it is 'a dynamic life form maintained over time by an informed genetic flow. The individual represents (re-presents) a species in each new generation. It is the token of a type, and the type is more important than the token' (Rolston 1988, 143). Rescher (1980) finds an ethical duty to conserve species on the grounds that species possess value, and that we have a general obligation to protect value.

Ethics have also been developed which consider greater wholes of which species form a part. The most influential ethic of this kind is Aldo Leopold's famous Land Ethic. This 'enlarges the boundaries of community to include soils, waters, plants, and animals, or collectively: the land' (1949, 202) and 'changes the role of *Homo sapiens* from conqueror of the land community to plain member and citizen of it . . . [and] implies respect for his fellow members, and also for the community as such' (1949, 204). At the heart of the Land Ethic is the maxim that

> A thing is right when it tends to preserve the integrity, stability, and beauty of the biotic community. It is wrong when it tends otherwise.

Tansley (1935) replaced the term 'biotic community' with the term 'ecosystem', which has now gained general usage. Thus, as reformulated by James Heffernan (1982), the land ethic is:

> A thing is right when it tends to preserve the characteristic diversity and stability of an ecosystem (or the biosphere).

Callicott (1980, 1994) and Devall and Sessions (1984) make the point that our obligations are not only to individual animals, but also, indeed principally, to species and the biotic community as a whole. Gunn (1980) proposes that our concern for rare species is best dealt with through an environmental ethic which affords intrinsic value to each species, and also to ecological wholes. Weir (1989) combines Kantian duties to rational animals (especially humans) with a duty to preserve species, populations and ecosystems. The ultimate extension of this line of ethical reasoning is reached in James Lovelock's thought-provoking 'Gaia hypothesis' which maintains that the earth is one whole living organism, with discernible organs (e.g. the tropics and the continental sea areas) which must be protected (Lovelock 1979).

As with the other major schools of environmental ethic, we can note a number of possible weaknesses in ecocentric positions. Deep Ecology has been identified as lacking in concerns of social and intra-human justice, as well as focusing on 'pristine wilderness'; as such it may be inappropriate for European and developing countries (Guha 1989). As we saw in Chapter 2, there is controversy over the very existence of wholes such as species (rather than simply the aggregation of individual organisms). Furthermore, even if ecocentric ethics do provide a basis for the moral considerability of 'wholes' – which not all accept (e.g. Cahen 1988) – they may still have difficulty in dealing with conflicts *between* wholes, or between wholes and individual components (i.e. individual animals or plants). As far as the wholes–wholes tension is concerned, some form of meta-ethic may be necessary to guide actions when the choice is between protection of a number of competing ecosystems or species. Should we, for instance, allow water abstraction, if such abstraction will tip the balance from preservation of an aquatic ecosystem in favour of a new dry-land community? Are there any good reasons to prefer aquatic ecosystems over dry-land ecosystems, or pre-existing ecosystems over ecosystems that result from human intervention?

There is also difficulty in relations between wholes and individual plants or animals. The preservation of certain species – e.g. *Plasmodium vivax*, the malaria parasite – may be contrary to the interests of individual humans. Caring for species or ecosystems may require the whole to take precedence over the parts, lending weight to accusations of 'environmental fascism' (Regan 1983) or 'totalitarianism' (Kheel 1985). A practical working out of the ethical tension between wholes such as species and individuals occurs when it is proposed that the life or liberty of some animals be sacrificed for the greater good of a population or group. Sometimes ecosystem/species protection seems to require the extermination of individuals of some other invasive species (e.g. extermination of escaped mink in order to protect the British wild vole population as required by the 1985 Eradication of Mink Bill).

Ecocentric ethics try to respond to accusations of ecofacism by acknowledging that individuals matter too (Marietta 1988). Yet it is difficult to avoid the conclusion that when the interests of wholes and individuals collide, ecocentric ethics favour the former over the latter.

Pluralistic ethics

It may be that each of the ethical positions briefly explored above has its merits, and that we need to use each in an appropriate context. Academic purity is all very well but in reality even those of us who accept the need for ethics tend to an element of 'mix and match' or 'context-specific application'. Pluralism is certainly not out of the question for human ethics: we may be utilitarian in agreeing to medical triage or social programmes, Kantian in urging abstinence from lying, rights-based in resisting erosion of civil liberties and human rights, and virtue-based in espousing the need for attitudes of gender respect and racial tolerance. It is possible that a similar sophistication is required in our moral dealings with nature. We may need to find a place for anthropocentric, biocentric and ecocentric values and simply accept the existence of apparent inconsistencies 'at the edges' (Stone 1985).

The ethical content of environmental law

Environmental ethics in international environmental law

It has been argued that the concerns of international environmental law were originally narrowly anthropocentric but, over time, have widened to a biocentric perspective (Emmenegger and Tschentscher 1994). Certainly anthropocentric instrumental references to the environment are not hard to find in many older international environmental instruments. In the case of the Declaration for the Protection of Birds Useful to Agriculture 1875, the title speaks for itself. Similarly, the preamble to the International Convention for the Regulation of Whaling 1946 refers to 'the interest of the nations of the world in safeguarding for future generations the great natural resources represented by the whalestocks'; terminology which, when applied to great whales, is ethically offensive from the biocentric perspective. Another example of fairly solid anthropocentric attitude is the Convention on Wetlands of International Importance Especially as Waterfowl Habitat (the 'Ramsar Convention'). This fairly weakly worded measure requires party states to ensure, as far as possible, the 'wise use' of wetlands. Although the convention itself did not define the term, the 1987 meeting of the parties agreed that 'wise use' means the 'sustainable utilization for the benefit of humankind in a way compatible with the maintenance of the natural properties of the ecosystem' and, in turn, explained that '*sustainable utilization*' means 'human use of a wetland so that it may yield the greatest continuous benefit to present generations while maintaining its potential to meet the needs and aspirations of future generations'.

Although human goods and interests have not ceased to predominate there are signs of increasing 'extension' of the ethical concerns in international law. One indicator is the now commonplace reference to duties to future generations (e.g. Biological Diversity Convention 1992, preamble; Berne Convention on European Wildlife and Natural Habitats 1979, preamble, Art. 2; Bonn Convention on the Conservation of Migratory Species of Wild Animals 1979, preamble; Stockholm Declaration 1972, Principles 2 and 6). Another is the occasional explicit recognition of the rights and welfare of individual animals (e.g. the European Convention for the Protection of Animals During International Transport, 13 December 1968).

D'Amato and Chopra (1991) attempt to show that the law relating to whaling has evolved from a strictly anthropocentric position, in which whales were conserved under the 1946 Convention for the Regulation of Whaling only because of the human interest in maintaining 'stocks' for future 'harvest', to a moratorium on hunting which amounts, in essence, to recognition of a 'right to life'. This putative right to life, it must be noted, is now under threat due to the ongoing refusal of whaling nations (Norway, Japan and Iceland) to relinquish whaling and the recent proposal to resume legal whaling under the 1946 Convention.

The move beyond biocentric to ecocentric approaches is evidenced by several of the more important late-twentieth-century environmental conventions. Many such conventions assert, in their preambles, that nature/species/ecosystems have intrinsic value (e.g. Biological Diversity Convention 1992, preamble; World Charter for Nature 1982; Berne Convention on the Conservation of European Wildlife and Natural Habitats 1979). Many are also ecocentric in substance. The 1973 Convention on International Trade in Endangered Species is a case in point: its main provisions require party states to prohibit the trade in specimens of species 'threatened with extinction which are or may be affected by trade' (i.e. listed in Annexe I) unless certain stringent criteria are satisfied. These include the requirement that the Scientific Authority of the State of export has advised that export of the specimen will not be detrimental to the survival of that species. Although the convention in practice is not particularly effective (Matthews 1996), and the practice of banning trade may do more harm than good by encouraging poaching and by removing the incentive for conservation (Favre 1993). Nevertheless, the attitude that the convention manifests is clearly one of respect for, and protection of, species.

One could be forgiven for thinking that the nearest approach to Leopold's Land Ethic in international law would be the 1992 UN Convention on Biological Diversity. After all, biological diversity is, as defined by the Convention:

> the variability among living organisms from all sources including, inter alia, terrestrial, marine and other aquatic ecosystems and the ecological complexes of which they are part; this includes diversity within species, between species and of ecosystems.

This concern for the wholes of species, ecosystems and their constituent variability (or 'richness') is the very thing that Deep Ecologists claim ethical priority for. However, the preamble to the 1992 convention reveals mixed reasons for wishing to preserve this diversity. In the preamble the parties to the convention profess their awareness of 'the intrinsic value of biological diversity' as well as 'the importance of biological diversity for evolution and for maintaining life sustaining systems of the biosphere'. The parties also, however, refer to many more human-focused reasons for biodiversity conservation including:

> 'meeting the food, health and other needs of the growing world population', to 'strengthen friendly relations among States and contribute to peace for humankind' and the need to 'conserve and sustainably use biological diversity for the benefit of present and future [human] generations, . . .

Thus, despite Emmenegger and Tschentscher's claimed shift in the ethical basis of international environmental law towards the biocentric perspective, the facts often point to a much more complex picture in which a variety of ethical positions can be discerned in most international environmental instruments.

Environmental ethics in conservation law

Conservation is the activity in which ethical presumptions about the value of nature, and its component parts, most readily become visible. This is due to the fact that, however it is arranged, conservation requires trade-offs to be made between human interests and the interests of nature.

Conservation may require additional *intra-nature* considerations such as how to balance the interests of individual animals and plants, on the one hand, with the interests of species and biotic communities on the other.

In the pre-industrial period in England and other European countries conservation was provided for by the feudal system of land ownership, and the ties of fealty that existed between lord and serf. In England, for example, land ownership was based around a life estate encumbered by the common law 'Doctrine of Waste'. The Doctrine of Waste required the life tenant to avoid inflicting any relatively permanent injuries on the land such as the cutting down of *timber* (i.e. oak, ash and elm of twenty years age or more – *Honeywood* v. *Honeywood* (1874)) or the extraction of minerals. The interests of the landlord and life tenant were balanced by the right of the life tenant to take *wood* (i.e. young trees) and through the concept of 'estovers'. The estovers of *house-bote*, *plough-bote* and *hay-bote* permitted the life tenant to take limited amounts of timber for fuel or house repairs, repair of agricultural implements and for fixing fences respectively. Other, more general positive acts of destruction – such as grubbing up woods, or from cutting saplings of insufficient growth or at unreasonable times – were prohibited under the category of 'equitable waste'. These legal rights resulted in a long period of sustainable management of woods, especially through the practice of coppicing (Rackham 1976). These early laws were, in essence, instruments for sustainable management. As such, consciously or unconsciously, they took account of the requirements of intergenerational equity. They were, however, decidedly anthropocentric – the idea of nature's instrinsic value, although known to the Greeks, had been lost from view by the Middle Ages (Thomas 1984).

UK Conservation legislation in the twentieth century began with the provisions of the National Parks and Access to the Countryside Act 1949 – now reproduced with modifications in the Wildlife and Countryside Act 1981. The report of the Wildlife Conservation Special Committee, on which the 1949 Act was based, expressed the view that,

> The concept of nature conservation, broadly interpreted, embraces several more or less distinct purposes . . . biological survey and research; experiment; education; and amenity.

Chief amongst these human-focused concerns was scientific study, with the system of National Nature Reserves (NNRs) and Sites of Special Scientific Interest (SSSIs) envisaged as 'living laboratories' (Evans 1992, 7). In keeping with its essentially anthropocentric approach, the 1949 Act introduced a representative system of nature protection, relying on small sites and reserves, oases in seas of urban and agricultural development. The Special Committee's report likened the proposed system of SSSIs and NNRs to a national museum, implying the need to include only token or representative examples of the range of habitats and species in England. It also endorsed the view that the aim of the ensuing legislation should be to give the most balanced result possible within the smallest overall area of land. Unfortunately, the SSSI/NNR conservation system contained inadequacies that allow landowners to avoid conservation restrictions. The most important of these was the ability of a landowner to give notice to the Nature Conservancy Council, and then wait for four months, and then be free from the legal restrictions on development. As LORD MUSTILL said, in *Southern Water Authority* v. *Nature Conservancy Council* [1992],

> It only needs a moment to see that this regime is toothless, for it demands no more from the owner or occupier of an SSSI than a little patience. Unless the council can convince the Secretary of State that the site is of sufficient national importance to justify an order under s.29 – as we have seen, a task rarely accomplished – the owner will within months be free to disregard the notification and carry out the proscribed operations, no matter what the cost to the flora on the site. In truth the Act does no more in the great majority of cases than give the council a breathing space within which to apply moral pressure, with a view to persuading the owner or occupier to make a voluntary agreement.

The advent of EU conservation laws, especially the Habitats Directive 92/43, brought about some fortunate strengthening of the system, although only in relation to the minority of sites covered by these Directives. The UK system has also been strengthened recently through the Rights of Way and Access to the Countryside Act 2000. Provisions of the 2000 Act enable the conservation agencies to impose permanent restrictions in place of the previous temporary four-month restrictions to prevent damaging operations on SSSIs in England and Wales. There is also a power for the conservation agencies to secure the active *management* of a SSSI (rather than a mere passive obligation to avoid harm).

In the United States conservation was, for the first half of the twentieth century at least, perceived in strictly utilitarian human-centred terms (Hays 1969). Forestry, in particular, was heavily influenced by the 'nature is for humans' ideas of Gifford Pinchot (Des Jardins 1997, 45–53).

Pinchot argued for 'an orderly scheme for national efficiency, based on the elimination of waste, and directed toward the best use of all we have for the greatest number [of humans] for the longest time' (1914, 23). Pinchot's 'wise use' conservation movement ousted the more ecologically sensitive views of John Muir, founder of the Sierra Club, who had argued that nature should be preserved for its own sake or (in his later writings) for spiritual replenishment and aesthetic appreciation (Nash 1989, 40). But the arrival of ecological awareness in the 1960s resulted in serious public doubt that the treatment of old growth forests as if they were mere crops was, after all, to the 'best use', as well as rejection of the utilitarianism of 'the greatest number for the longest time'. The interests of plants, animals, species and ecological units needed to be brought into the equation.

The single most important step in this shift towards ecocentrism was the 1973 Endangered Species Act (ESA) (16 USC 1531–1543). The main justification for the Act was that

> various species of fish, wildlife, and plants in the United States have been rendered extinct as a consequence of economic growth and development untempered by adequate concern and conservation; [and that endangered] species of fish, wildlife, and plants are of aesthetic, ecological, educational, historical, recreational, and scientific value to the Nation and its people.

The Secretary of the Interior is required to list endangered species and the critical habitats on which such species depend. These listing functions are delegated to the Fish and Wildlife Service and the National Marine Fisheries Service. Listing of endangered species must be based solely on scientific information; that is, economic considerations cannot be taken into account (Section 1533(b)(1)(A)). Listing of critical habitat is more discretionary and can include economic factors (Section 1533(2)).

The substance of the Act is a prohibition of the 'taking' of any species listed as 'endangered'. A species is 'taken' when it is 'harassed, harmed, pursued, hunted, shot, wounded, killed, trapped, captured, or collected'. Public authorities are required, by Section 7 of the Act, ' to ensure that actions authorized, funded or carried out by them do not jeopardize the continued existence of' an endangered species or 'result in the destruction or modification of habitat of such species'.

A key issue in the interpretation of the Act is the permissibility of balancing the values of nature conservation, on the one hand, against the human welfare derived from projects on the other. In *Sierra Club* v.

Froehlke (1976) the Court of Appeals allowed the Corps of Engineers to proceed with a project likely to imperil the Indiana bat, on the grounds that the Corps had balanced the expected benefits of the project with the 'importance of an unspoiled environment'. This approach was rejected in *Tennessee Valley Authority* v. *Hill* (1978). The case stemmed from the discovery of 'the snail darter' – a unique and interesting species, but one without any apparent human use value – in the Little Tennessee River towards the completion stage of a project to construct the huge Tellico Dam on that river. The Sierra Club sought to have the project halted. Federal authorities were understandably reluctant to do this since the (human) costs of abandoning the dam project would be massive. The snail darter, on the other hand, whilst thought to be unique to that river, was one of around 130 species of darters. The District Court declined to halt the project, taking account of the continuation of congressional funding even after discovery of the fish, as well as the fact that the dam was very near completion.

The case was perceived as a battle of environmental versus human interests. In the event ethical reasoning was only indirectly relevant. The question before the Supreme Court was not, 'is this species worth protecting' but, rather, 'did Congress, in passing the ESA, intend preservation of endangered species *at all costs*?' The legislative history of the Act, the court found, 'reveals a conscious decision by Congress to give endangered species priority over the "primary missions" of federal agencies'. Having determined the existence of 'an irreconcilable conflict between the operation of the Tellico Dam and the explicit provisions of section 7 of the [ESA]' the Supreme Court concluded that it was required to shut the project down.

The Tellico Dam decision created an ethical and political backlash: sections of the American public were outraged that a principle should require the sacrifice of a nearly completed project, worth around $100 million, for the sake of a small and apparently worthless fish. The ESA was amended, creating a special committee empowered to override the protection given to a particular species if it determines that there are no reasonable alternatives to the agency action, that the action is in the public interest, and the benefits clearly outweigh those of compliance with the Act. The 'God Squad', as it became known, refused to make such a determination for the Tellico Dam itself, but ultimately the dam was completed by passage of special enabling legislation (Note 1979). Despite the reluctance of the Committee to use its powers, the final position is that anthropocentric considerations have been reintroduced into ESA decisions in order to override its essentially ecocentric basis. The final irony for the 'snail darter' is that the fish was discovered in several other locations, so the Tellico Dam project would not, after all, have threatened its existence.

The situation in Europe shows some equivalence in the evolution of the interplay of law and ethics for species protection. In 1979 the European Community adopted Directive 79/409/EEC for the protection of Wild Birds. Article 4 of this required Member States to 'classify in particular the most suitable territories in number and size as *special protection areas* (SPA) for the conservation of [endangered bird species listed in Annexe I to the Directive]'. In a number of cases the European Court of Justice considered the question at the heart of *Tennessee Valley Authority* v. *Hill*; that is, whether there could be a trade off between economic and social benefits, on the one hand, and the duty to protect endangered species on the other. *Commission* v. *Germany* [1991] concerned a dyke in the Ostfriesische Wattenmeer, an area situated in Niedersachsen by the North Sea, being a wetland of international importance under the Ramsar Convention and also an SPA under Directive 79/409. In 1985 German authorities authorised modification of the dyke as part of a coastal defence works programme. However, in so doing, they went beyond mere coastal defence and enlarged the dyke in order to facilitate access of fishing vessels to the Port of Greetsiel. The Commission argued that this was damage to an SPA and, as such, amounted to a breach of Article 4 of the Directive. The ECJ ruled that

> the power of the Member States to reduce the extent of a special protection area can be justified only on exceptional grounds [which] . . . must correspond to a general interest . . . superior to the general interest represented by the ecological objective of the directive. In that context . . . economic and recreational requirements, do not enter into consideration.

In other words the ECJ would not allow economic and social benefits to override the interests of the species dependent upon the habitats protected by Article 4 of the Directive. In *R.* v. *Secretary of State for the Environment, ex parte Royal Society for the Protection of Birds* the ECJ was asked to give an interpretative ruling on the question of whether a Member State could exclude an area which would, on ornithological criteria, be part of an SPA, in light of economic and social considerations. Again the ECJ answered in the negative. Many EC Member States were unhappy with this uncompromising prioritisation of species and ecosystem interests. During negotiation of a broader conservation measure – the Habitats Directive – the opportunity was taken to water down the obligations. Under the new Directive, endangered species and habitats are viewed holistically as forming an overall European nature resource ('Natura 2000'). Ecosystems which are themselves endangered or which are habitat to endangered species are required to be protected as 'Special Areas of Conservation' (SACs). However, unlike SPAs under the Birds Directive, SACs may be modified in certain circumstances. These are that (1) there is no alternative to the project concerned, (2) the project must be carried out 'for imperative reasons of overriding public interest, including those of a social or economic nature', and (3) the Member State takes 'all compensatory measures necessary to ensure that the overall coherence of Natura 2000 is protected'.

The idea that compensation is due if ecological interests are infringed was, you will recall, part of Paul Taylor's theory of 'respect for nature': specifically, the 'restitutive duty'. Taylor's duty operates at the level of the individual organism, which is problematic: how can improving the lot of surviving sea-birds amount to compensation to those birds killed in an oil spill? In the EU system, the compensation ethic is more tenable. If the object to be protected is European nature as a whole then protecting or restoring some habitat to make good the loss of part of an SAC makes sense. A law that awards or demands restitution when ecosystem interests are damaged is in keeping with ecocentric ethics (Baker 1995).

Ethics, law and hunting

One of the major divisions between biocentric ethics, on the one hand, and anthropocentric and ecocentric ethics, on the other, is their attitude towards individual animals. Theories of animal rights or welfare generally require abstention from animal killing, especially hunting (i.e. killing that is both unnecessary for human life and from which pleasure is obtained).

Sometimes ethical concern for species can result in almost 'accidental' protection of individual animals. In *Defenders of Wildlife* v. *Andrus* (1977) the NGO sued the US Department of the Interior, seeking to force it to prohibit dusk and dawn shooting of ducks, on the grounds that at these times more ducks belonging to endangered species are accidentally shot. The court thought that the impact of dusk and dawn shooting might be considerable and instructed the Interior Department to adjust the hours during which it legally permitted hunting. This case evinces no concern for ducks *as ducks* – only a determination of the tension between human

Box 8.1

Old growth forests: an 'ethics and law' conflict in the United States

The Pacific Northwest region of the United States contains the most significant reserves of 'old growth' forests left in the country. These happen to be the habitat of the northern spotted owl, *strix occidentalis caurina*. The Federal Wildlife Service's (FWS) refusal to list the owl as an endangered species was successfully challenged, and overturned, in the *Northern Spotted Owl Decision* (1988). The owl was finally listed as an endangered species from 23 July 1990. Once listed, attention turned to the owl's 'critical habitat'. As previously noted, listing of critical habitat is more discretionary than listing of species. The Secretary of the Interior is required to list habitat 'to the maximum extent prudent and determinable' for each listed endangered species, but may exclude areas if the benefits of exclusion outweigh the benefits of inclusion (Section 1533(b)(2)). The *Thomas Report* (1990) recommended protection of owls by designation of large habitat blocks – 'Habitat Conservation Areas' (HCAs) – containing twenty or more pairs of owls. The report also recommended cessation of logging activities in HCAs, and other conservation measures such as discouraging road building, elimination of clear-cutting, and ongoing monitoring. The FWS, realising that such measures would be deeply unpopular, declined to list critical habitat for the owl, ostensibly on the grounds that the extent of habitat required could not be determined. In February 1991 the federal district court ruled that this refusal to designate was unlawful. Following further litigation by the Audubon Society to halt the sale of any further logging rights, the FWS eventually listed 6.9 million acres in Oregon, Washington, and California as critical habitat for the owls.

The ethical, political and legal battle over the northern spotted owl is symptomatic of a general conflict at the heart of the ESA between human economic interests and conservation values. The problem is difficult to resolve because of the high levels of discretion given to the FWS for the listing of habitats for endangered species. If old growth forests and owls are both to be protected then, Meyers concludes, the US conservation legislation needs to be amended. Specifically, a duty ought to be included requiring federal agencies to 'reject project alternatives whose benefits to [human beings] are outweighed by the adverse impacts on the "whole community"' (Meyers 1991, 662). In other words, ecological ethical priority requires to be 'fixed' in law, not left to agency discretion.

The nexus between the bird and its habitat is an interesting example of value reversal: the bird, which is the legal object of preservation, has become the instrument through which the habitat is preserved. Although every school of environmental ethics could, potentially, support the preservation of this rare owl, its proper ethical locus is at the level of species-based ethics. On the other hand protection of old growth forests is most comfortably accommodated within ecocentric environmental ethics. Meyers (1991, 625), who finds this reverse instrumentalism less than ideal, points out that, 'We ought to be able to protect old-growth forests because of their inherent worth and ecological value, not merely because they provide living space for northern spotted owls and pacific yews.' Baker (1995, 707), in similar vein, argues that 'the law misdirects its energy if it conceptualizes the forest in terms of its constituent parts'.

pleasure and the interests of endangered species. Similarly, in *Organised Fishermen of Florida* v. *Andrus* (1980) a decision to restrict fishing in the Everglades National Park was challenged. The basis of the restriction was the protection of crocodiles – creatures that are sensitive to human disturbances, especially from motorboats. Even though the decision would allegedly result in the loss of commercial fishing worth about $1.2 million annually, the court upheld the decision on the grounds that the species would benefit from the restriction. Of course, many thousands of individual fish benefited incidentally from the desire to protect rare crocodile species.

The interests of species, ecosystems and humans, on the other hand, do not necessarily require such proscriptions, and may on occasion even require hunting (or at least capture) to take place. A few cases illustrate the ethical dimensions.

In the New Zealand case, *The Kaimanawa Wild Horse Preservation Society Inc.* v. *Attorney-General* [1997], the NGO challenged a plan, approved by the Minister of Conservation, to cull wild horses which roam tussock land in the south-western Kaimanawa mountains in the central North Island. The herd had grown in size until it threatened to interfere with 'endangered, rare, and biogeographically significant plants, and on ecosystems of tussock grasslands, subalpine herbfields, wetlands and forest margins'. The NGO argued that the proposed cull would compromise (if not destroy) the horses' unique genetic traits, and would defeat the ability of the horses to continue the natural selection processes which had made them so distinctive. The NGO's legal claim was that the cull plan infringed the duty of environmental protection applicable to every person by Section 17 of the 1991 Resource Management Act. Specifically Section 17 requires a person 'to avoid, remedy, or mitigate any adverse effect on the environment arising from an activity carried on by or on behalf of that person, whether or not the activity is in accordance with a rule in a plan, a resource consent, . . .'

The case contained a clear ethical conflict between the value of protecting a species (or, more accurately, a population) versus the value of protecting an ecosystem. Judge Sheppard, perhaps not unsurprisingly, ducked this issue, instead resolving the application by reference to the statutory interpretation of Section 17. First, he concluded that the words 'activity' and 'environment' were to be given very broad definitions. Unless qualified in some way this would lead to the (impliedly absurd) situation where persons could be restrained from a wide range of activities including controlling possums, clearing scrub and weeds, or engaging in personal behaviours that had an adverse effect on other people. To avoid this conclusion Judge Sheppard reasoned that the Section 17 duty applies only to activities of the kind controlled by the other provisions of Part III of the Act (i.e. activities of a 'purely' environmental nature; that is, pollution prevention, habitat conservation). Although Part III includes 'the use of land' the cull plan did not, in his opinion, fall within this term.

National Audubon Society v. *Hester* concerned a challenge to a decision to take the last wild condors into captivity, in order to breed and release more in an effort to save the species. By the late 1980s the California condor, the largest North American bird, had declined to the point where only twenty-six members of the species remained. Of these, twenty were in zoos in Los Angeles and San Diego as part of a breeding programme designed to avert extinction of the species. Only six wild birds were still in existence. The FWS initially decided to capture only those birds whose genes were poorly represented among the captive flock, thus maintaining a small wild flock, with the intention of eventual release of young birds bred in captivity. Maintenance of some wild birds was seen as important to act as 'guide birds' for the captive-bred condors ultimately to be released. The FWS later reviewed and reversed this decision, deciding to bring in all remaining condors, after deaths from lead poisoning led to concerns for the safety of the remaining wild individuals.

The National Audubon Society requested and obtained an injunction barring the Service from carrying out that decision. The Appeals court, however, reversed the injunction on the grounds that the agency's decision constituted a reasoned exercise of its discretion in fulfilling its statutory mandate. Under the ESA, an agency's determination that its action will not threaten endangered species is to be set aside only if arbitrary and capricious. The FWS's decision was defensible and not manifestly wrong, since the potential harm of leaving or bringing in the birds was finely balanced.

Several environmental ethics were in conflict in this case. The Audubon Society was ostensibly acting in what it considered was the best interests of the surviving wild condors. For the condors in question the 'bring in' decision would be the end of their wild lives and, as such, could be viewed as a gross infringement of their right to live in a state unmolested by humans (i.e. Taylor's duty of non-interference). On another view the Society was more interested in protecting the anthropocentric interests of its members who sought to retain the aesthetic pleasure of watching the wild birds in their natural environment. Then there was also the conflict between the ethics of species preservation and the ethics of individual respect. Would it be right, ethically, to imprison the last few humans in order to preserve the human race? Was it ethically justifiable to infringe the liberties of the individual condors for the sake of the continuation of their species line? Clearly the FWS thought so, and I suspect that most people would agree.

A notable aspect of the *Fewings case*, discussed in Box 8.2, is the court's unwillingness to enter into any kind of ethical reasoning on its own behalf. As the trial judge, Laws J, commented:

> The court has no role whatever as an arbiter between those who condemn hunting as barbaric and cruel and those who support it as a traditional country sport . . . This is of course a question on which most people hold views one way or the other. But our personal views are wholly irrelevant to the drier and more technical question which the court is obliged to answer. That is whether the County Council acted lawfully in making the decision . . .

This firm rejection of a role as an ethical adjudicator is also witnessed in US courts. In *Tennessee Valley Authority* v. *Hill* (1978), for example, Burger CJ, delivering the opinion of the Court, strongly resisted the suggestion that federal courts were empowered by the ESA to make 'fine utilitarian calculations' or to 'pre-empt congressional action by judicially decreeing what accords with "commonsense and the public weal"'.

Environmental ethics and pollution law

Pollution legislation has the potential to be a positive force for the upholding of ethical environmental values. Since pollution is a threat to

Box 8.2

Hunting in the English countryside: judicial ambivalence towards ethical motivations and the ethics of direct action

Hunting with hounds – for fox, otter and deer – has a long and culturally important history in English society. In the last few decades, however, opposition to this pursuit from the anti-hunting lobby has grown significantly.

In *R.* v. *Somerset County Council, ex parte Fewings and others* [1995] the applicants, representing the Quantock Stag Hounds, sought judicial review of Somerset County Council's decision to ban deer hunting on their land. The applicants argued that in basing their decision on the alleged cruelty of deer hunting the Council had acted unlawfully.

The Council's decision to ban hunting was governed by Section 120(1)(b) of the Local Government Act 1972 that empowers local authorities to acquire and manage land for the 'benefit, improvement or development of their area'. The trial judge, Laws J concluded that:

> Section 120(1)(b) [of the 1972 Act] confers no entitlement on a local authority to impose its opinions about the morals of hunting on the neighbourhood. In the present state of the law those opinions, however sincerely felt, have their proper place only in the private conscience of those who entertain them. The council has been given no authority by Parliament to translate such views into public action.

The applicants appealed to the Court of Appeal (1995) which, whilst upholding the lower court's decision, nevertheless took a broader and more sympathetic view of the relevance of ethical considerations. According to the majority of the Court of Appeal judges (Sir Thomas Bingham MR and Simon Brown LJ), it could not be said that considerations of cruelty were irrelevant in determining what was for the 'benefit of the area'. Simon Brown LJ went further and commented that,

> the cruelty argument, as well as the countervailing ethical considerations, were necessarily relevant to the decision. Had they been ignored I believe that the council would have been open to criticism.

Even if the council had been able, lawfully, to take into account the cruelty arguments, this would not have been the only relevant consideration of an ethical nature. A report prepared by the County Planning Officer prior to the hunting ban decision expressed concern for the deer herds and local population if hunting were forced to cease. Not only would some other form of population control be necessary (e.g. shooting) but the cooperation of neighbouring landowners, which would be necessary for the success of any such alternative management programme, might not be forthcoming. The report commented:

> it is the existence of the hunt which leads many local farmers and landowners to tolerate crop damage from the deer . . . The majority of local farmers are sympathetic towards the deer. Many people maintain that the disappearance of hunting would eventually lead to the loss of that sympathy and a steady increase in shooting, possibly resulting in the extinction of the red deer in the hills.

It could be argued that, in such circumstances, the main duty of the Council is to consider all major ethical positions before reaching a decision, including anthropocentric, biocentric and ecocentric values.

the health and life of all forms of life, pollution legislation offers a mechanism for the concrete expression of a 'right to life', duties to individuals or even species and ecosystems. All of these require the absence of human-originated chemicals as a precondition for flourishing.

The common law allows the effects of pollution on animals, plants and ecological wholes to be taken into account indirectly in two ways: first, as part of the overall determination of whether a use of neighbouring land is unreasonable; second, as part of the calculation of damages due for that interference. Injury or death of wild animals and plants resulting from pollution will, if the causal agent is an 'unreasonable user', be subject to compensation. By what are, strictly speaking, anthropocentric considerations, the court may be able to take into account species or ecosystem disturbances. Thus, in *Marquis of Granby* v. *Bakewell UDC* (1923), in which water pollution destroyed a valuable natural trout fishery, the court awarded damages not only for restocking but also a sum for loss of 'wildness' of the restored fish.

Generally speaking, the more precise pollution standards contained in legislation are set by reference to their effects on *humans* or, at least, by human values. For instance, the National Ambient Air Quality Standards set under the US Clean Air Act must be adequate 'to protect *public* health' with 'a margin of safety'. Similarly, the water pollution standards in the early EU Water Directives were set, principally, by reference to their human health effects (ENDS 1992). Occasionally, legislators or the judiciary grasp the interrelated nature of conservation and pollution controls. For instance, in *Commission* v. *Spain* [1993] the ECJ concluded that the obligation, under Article 4 of the EC Wild Birds Directive, to identify and protect a Special Protection Area, requires positive action in the form of treatment of polluting discharges.

A further problem with integrating environmental ethics and pollution legislation is that the latter is underpinned by a philosophy of *technocentricism* – the optimistic belief that human progress and scientific knowledge can develop technological solutions to all environmental problems and that, therefore, key environmental decisions should be taken by technical or scientific 'experts' rather than the political expression of ethical standards (O'Riordan 1981; Bern 1990). Concrete environmental standards are reserved as matters for central government and/or environmental regulatory agencies (as for instance the process of formulating statutory water quality objectives under the Water Resources Act 1991).

Matters are made worse – from the point of accessing environmental values – by the fact that pollution legislation, in many developed states, includes requirement for cost–benefit analysis (CBA) or some other costs formula. The ethical poverty of using costs and benefits rather than *values* as criteria for environmental protection is itself well documented (e.g. Sagoff 1988; Wenz 1988). The central criticism of CBA methodology is that it uses monetary measures of consumer preference, such as 'willingness to pay', 'willingness to be compensated' and 'shadow pricing', to evaluate the acceptability of environmental damage or protection (Winpenny 1991). Surveys which obtain such measurements force respondents to behave towards the environment as 'consumers' rather than 'citizens', thereby replacing *values* with *preferences* (Sagoff 1988). The employment of such monetisation techniques

> confuses what people believe in and care about with what they desire and will spend money on . . . Suffice it to say that market analysis, when carried on in these terms, is a subversion of public debate. Economic analysis, carried on in these terms, can do nothing to reveal or clarify values other than those of economists themselves.
>
> (Sagoff 1981, 307, citing Bern 1990)

Research reveals that people reject the denigration of the environment which is intrinsic to this approach, and prefer the view that nature should be thought of as having rights to continued existence that are independent of the monetary costs/benefits involved (Hanley and Milne 1996).

The problem of CBA in pollution legislation is marked in the United States where obsession with efficiency has, in certain cases, undermined the substantive goals of environmental legislation. For example, the Clean Air Act of 1970 required the US EPA to set primary National Ambient Air Quality Standards adequate 'to protect public health' with 'a margin of safety'. This requirement was undermined by a later Executive Order prohibiting action by federal regulators except where the benefits exceed the cost.

A shift towards CBA is also noticeable in the UK. CBA was implicit in the requirement to adopt the Best Available Techniques *Not Entailing Excessive Cost* (BATNEEC) for processes which came within the Integrated Pollution Control regime established by the Environmental Protection Act 1990. The CBA requirement is more extensive in the Environment Act 1995, which establishes and governs the administration of the Environment Agency. Section 4(1) of the 1995 Act requires the Agency to take any likely costs into account in discharging its functions; Section 39 provides that, in considering whether to exercise any of its powers, the Agency shall, unless it is unreasonable for it to do so, take into account the likely costs and benefits of the exercise or non-exercise of that power.

Bern (1990) is concerned that pollution legislation,

> with its emphasis on 'rational' economic decision-making and technological solutions, is responsible for generating attitudes that prevent the individual from recognizing his or her role in causing and potentially reducing degradation.
>
> (Bern 1990)

Bern argues, using the US Clean Air Act rules as exemplary material, that existing anti-pollution laws not only foster an unwarranted faith in technological solutions, but that this faith in turn 'impedes the development of behaviour-based ethics'. Bern's argument has some substance and he is not the first to observe that reliance on law itself can have the unintended consequence of reducing the sense of personal responsibility for one's own actions (e.g. Schweitzer 1923). Just as criminalisation of an act signals its moral unacceptability, so too the legalisation of pollution through systems of permits sends out signals of acceptability. Emond (1984), in this respect, has observed that,

> Environmental protection legislation . . . is misnamed: in the intent, protection is quite secondary. The desire to facilitate development by keeping environmental degradation within 'tolerable' limits – usually expressed as 'maximum permissible levels' of contaminants – is paramount. The legislation is utilitarian not utopian. It lacks vision. Pollution is rationalised and, after the necessary permit is issued, legalised.
>
> (Emond 1984, 340)

Conclusion

Freyfogle (1994) argues that American federal environmental laws

> simply do not form a coherent moral order, nor do they convey a premeditated vision of ecological well-being. Because of this failure, environmental law-making is approaching a crisis of vision and imagination, stumbling on such knotty issues as nonpoint-source water pollution and declining wildlife habitat.
>
> (Freyfogle 1994, 819)

This lack of coherence is, to an extent, evident in environmental ethics and environmental law in every forum in which that nexus is encountered. The question is whether the lack of a coherent moral order is a good or a bad thing. If we agree with Sagoff that *values* should be determined by open public debate, and if we accept that society is likely to be *pluralistic* in its values towards the environment, then maybe this is not, after all, a reason for pessimism.

Key points

- Environmental ethics fall into several major schools or divisions. The normative consequences for any given environmental protection issue may vary widely according to the ethic that is adopted.
- Environmental law as a whole has shown a steady movement towards ecocentric ethical positions. On the whole, however, it is still largely underpinned by unstated anthropocentric assumptions and values.
- Courts usually wish to maintain ethical neutrality. They often find, therefore, mechanisms for deciding cases on 'technical grounds', leaving the value conflicts to one side.
- It is not clear how far or in what circumstances public bodies may take publicly held ethical environmental values into consideration within their legal mandate for action.

Further reading

Alder, J. and Wilkinson, D. (1999) *Environmental Law and Ethics*, London: Macmillan, and Gillespie, A. (1997) *International Environment Law, Policy and Ethics*, Oxford: Oxford University Press, provide broad-ranging discussion of the legal–ethical relationship.

D'Amato, A. (1990) 'Do We Owe a Duty to Future Generations to Preserve the Global Environment?', *American Journal of International Law*, vol. 84, no. 1, pp. 190–8. Somewhat dated but still a good examination of the merits and problems of legalising the future generations ethic.

Des Jardins, J.R. (1997) *Environmental Ethics: An Introduction to Environmental Philosophy* (2nd edn), Belmont, Calif.: Wadsworth, offers a clear and well-structured account of environmental ethics, using lots of interesting case studies and, in some places, legal examples.

Emmenegger, S. and Tschentscher, A. (1994) 'Taking Nature's Rights Seriously: The Long Way to Biocentrism in Environmental Law', *Georgetown International Environmental Law Review*, 6, 545–92. The authors provide numerous legal examples to argue that environmental law is making a 'paradigm shift' towards ecocentrism.

Discussion questions

1. Which kind of environment ethic (or ethics) do you find plausible as a basis for preventing environmental harm? How do you deal with the criticisms of your ethical foundation?
2. All environmental ethics seem to have some 'crunch point' at which the interests of humans are given priority over those of other members of the biological community. Where do you draw the line and how do you justify it?
3. Is ethical pluralism useful as a way of thinking about conserving the diverse range of natural entities that exist? If so how can we decide which ethical system should be applied to which environmental entity?
4. If the *Kaimanawa Wild Horse* case, discussed on pp. 239–40, had been decided by reference to the competing ethical values concerned rather than narrow grounds of statutory interpretation, what ought the outcome to have been?
5. Should the law require public decision-makers to take account of publicly held values concerning the environment? Does this offer a route to the infusion of environmental ethics into the law, or is it a dangerous step towards confusion of ethics with populist prejudices?

9 Politics, law and the environment

- Capitalism and technocentricism
- Socialism
- Anarchism
- Liberalism
- Democracy

The focus of this chapter is on the relationship between law and politics in the quest for environmental protection. Political structures are, without doubt, amongst the most important features on the landscape of global governance. Effective political institutions carry the capacity to bring about change, and change is necessary if the trend of environmental degradation is to be slowed.

The questions that we shall be discussing in this investigation are:

- What kind of political systems are most conducive to environmental protection?
- Can strong environmental laws be accommodated within liberal political institutions?
- How can the growth of environmental law be accounted for by political theory?

Political systems, law and the environment

What kinds of political systems are well suited to the introduction of laws for environmental protection? This question is a subject for study in its own right that can only be touched upon here. However, from the standard literature in the area (e.g. Dobson 1990) two central issues can be ascertained. The first is that there is considerable disagreement about whether capitalist, socialist, or anarchistic polities are the most conducive to environmental protection. The second is that it is also unclear whether liberal/democratic or authoritarian type polities are better placed to achieve environmental protection.

In this section capitalism is taken to mean any political system that allows the economic means of production to lie in largely private hands, and that allows for the accumulation of capital to effect such a purpose.

Capitalist systems typically have certain key characteristics:

- favouring *technological solutions* to environmental issues;
- placing reliance on *economic growth* through *free trade* as a means of ameliorating environmental damage; and
- utilising *liberal/democratic institutions.*

In this section of the chapter these characteristics will be more fully stated and their legal connections explored.

Technological solutions

The preference, in capitalist economies, for technological solutions to environmental problems is typical of those whom O'Riordan (1981) describes as 'technocentrists'. Technocentrists are understood by their distinction from 'ecocentrists'. Ecocentrists regard nature as a friend, as a community that includes humans and all other living things as part of a larger whole. They often recommend small, self-contained communities, self-sufficiency, and frugal consumption in line with what they regard as natural patterns of behaviour (e.g. Schumacher 1973). Ecocentrists tend to distrust the free market as short-termist, wasteful of resources and the cause of unnecessary consumption. Many ecocentrics are also deeply techno-sceptic, and argue that reliance on technology is the cause of most environmental and social problems (e.g. Unabomber 1997).

Technocentrists, on the other hand, generally regard nature as requiring human domination or, at least, human management. Optimistic technocentrists believe that technological innovations can solve any environmental problem. North (1995) is typical of this position:

> the wellbeing of humankind, and the planet on which we live, depends on technical sophistication rather than Ludditism. The rich world has spawned a good deal of doubt and fear about its own progress, and affects a dislike of its own technology. Many people fantasise that there is some sort of simple life which might be an improvement. This view is very flawed.
>
> (North 1995, 5)

prohibited outright. Instead, the general approach is to rely on the use of further technology to ameliorate the environmental problems or risks created by technology's original application. This trend is well illustrated by the frequent stipulation of the application of Best Available Control Technology (BACT) in US environmental law, or the Best Available Techniques (BAT) in the UK and Europe to heavy industry. Certain laws – e.g. the Montreal Protocol to the Vienna Convention on Ozone Depletion – specify that best technologies should be transferred to developing countries in order to assist these in dealing with environmental problems. The extreme degree to which environmental law is essentially *technology* law (i.e. a matter of selecting some ideal technical solution) arguably alienates the specialists in this field from the object of their concern.

One possible explanation for this technophilic characteristic is that technology and innovation imply new markets for the industries involved in their development, and techno-heavy systems favour the interests of large established corporations over those of smaller or new entrants (Wilkinson 1997). Larger, more powerful corporations are significant forces in the acceptance and shaping of environmental laws at every level.

A second technocentric tendency is that the common law, which we have seen provides much of the background law of environment protection, is typically highly receptive to changes in industrial and technical practice. These result in the *development* of property which, in turn, is so clearly endorsed by John Locke's 'labour theory'.

The requirement of nuisance law, that the uses of land do not cause an 'unreasonable interference' with neighbouring land, is not at all restrictive of particular practices. Indeed the 'locality doctrine' is designed to facilitate reception of new strategies and technological innovations, even if these increase environmental stresses on neighbouring citizens. The limitation of liability in all areas of common law civil liability to 'reasonably foreseeable damage' has a similar effect since, as history shows (e.g. nuclear power or CFCs), knowledge about the full negative environmental consequences of a technology is rarely available at the time of its original use.

Third, environmental law can be seen in a more general way as supportive of technology: regulation of industrial and scientific practices provides a form of legitimation that helps to assuage public fears about their safety and to entrench the relatively free hand of industrialists. To this extent environmental law may, in the long term, be a negative factor in the overall battle to protect nature.

In this context we may also ask whether *all laws* in existing capitalist economies are designed in such a way as to promote the steady march of technology. The legal system *as a whole* provides the ordered social conditions necessary for the domination of humanity. Corporate laws and anti-trust laws exist to provide the basic framework for entrepreneurship. Laws which exist to protect intellectual property (patents, copyright, etc.) protect corporate endeavours. Contract law allows orderly relations between the originators of technology. Anti-strike laws and employment laws allow a stabilised stream of labourers for the manufacturing process. Tort laws (e.g. negligence) soften the blow of the inevitable mishaps of the application of technology. Criminal laws and property rights protect items of technology purchased by consumers. There is a strong sense in which Law (with a capital L) serves technology and technocentric modes of life that capitalism both requires and encourages.

Economic growth and free trade

Belief in the necessity of economic growth is at the heart of capitalist political systems. The question of whether growth is desirable, or even possible in the long term, has already been discussed (see p. 175–8). It will be recalled that 'ecological modernisation' theory stipulates that growth is not only compatible with environmental protection but, indeed, is *necessary* for its achievement. Johnson and Beaulieu (1996), for instance, suggest that at about $5,000 per capita GDP – the effect of growth on the environment becomes positive; significant sums can then be diverted to environmental protection. On the other hand, ecological pessimists view continuous economic growth both as theoretically impossible and environmentally destructive. Sarkar (1999), for example, argues that not even stabilisation or a 'steady state economy' will suffice: what is required is *contraction* of all developed economies.

One of the most significant problems of economic growth is, as McLaughlin (1993, 21) notes, the increasing extent to which humans are 'biosphere people' rather than 'ecosystem people'. Ecosystem people, of whom indigenous peoples are a good example, live in close connection with the means of their own subsistence. Being dependent in this way entails very short feedback loops between human action and ecosystemic effects. It is, therefore, obvious to an ecosystem person when social activities are wrecking the underlying basis upon which she depends. The modern person living in a developed capitalist society is, on the contrary, usually significantly removed from the effects of her actions: food, clothes, raw materials, many consumer goods – all are imported from other states. The observable effects of the consumer lifestyle are ameliorated by their 'projection' to other points of origin. Biosphere people thus come to suffer from the delusion that their life practices are environmentally benign, even ecologically preferable to those of more 'primitive' people. Global trade regimes exacerbate this problem by increasing the tendency for geographical separation between production and consumption.

How is law linked to the economic growth aspects of capitalism? The most obvious connection is the fierce defence, by laws at every level, of free trade. Trade between different geographical regions is conducive to economic growth partly because of the phenomenon of 'natural advantage' and partly because trade is amongst the main engines of consumption which, in turn, drives the capitalist economic machine. As we have already seen, regional and global structures are based on the premise of free trade. The central ethos of the European Union, NAFTA and the WTO is economic growth through the furtherance of international trade. Consequently, legal protection of the environment can only be allowed in rather narrowly defined circumstances. Thus, in the Danish bottles case (*Commission* v. *Denmark* [1988]) the European Court of Justice limited the circumstances in which EU Member States can restrict trade in order to protect the environment. Environmental protection can only occur if the laws concerned are non-discriminatory between Member States, and are proportionate to their objectives; by which the ECJ really means moderate and not radical in their demands.

In order to obtain the resources that fuel the global economic system many developing countries are inclined to turn to practices that are not only environmentally damaging but which also severely prejudice the peaceful and environmentally benign practices of indigenous peoples. The case of *Beanal* v. *Freeport-Moran, Inc.* (1997) is instructive in this matter. Here, Beanal – leader of the Amungme Tribal Council of Lambaga Adat Suku Amungme (LEMASA) – alleged that Freeport – operator of an Indonesian copper, gold and silver mine – engaged in human rights and environmental abuses through its security guards 'in conjunction with third parties'. The alleged offences encompassed:

1 death and violence (including the fatal stabbing of one member of his group, killing after torture of three others and 'disappearance' of four more);

2 torture (including kicking with military boots, beating with fists, sticks, rifle butts, stones; starvation, standing with heavy weights on the subject's heads, shackling of thumbs, wrists and legs);
3 surveillance and destruction of property, inducing mental stress and fear;
4 environmental violations from various mining practices carried out in and nearby the locality where he resided (destruction, pollution, alteration, and contamination of natural waterways, as well as surface water and groundwater sources; deforestation; destruction and alteration of physical surroundings); and, through these,
5 the purposeful, deliberate, contrived and planned cultural demise of the Amungme culture.

Although the petitioner lost the case, due to technicalities of the laws concerned (see p. 60), his allegations were neither proved nor disproved. There are good reasons to believe that, on the whole, global economic growth induces significant environmental and human rights abuses.

The above factors indicate that because of its emphasis on unlimited private consumption, technological 'solutions' to technological problems, unlimited economic growth and free trade, capitalism will never be able to provide an effective basis for robust environmental protection (Sarkar 1999; McLaughlin 1993). On this view it is unlikely that the laws drafted within that type of regime that purport to prevent environmental degradation can do more than slow the rate of damage, or provide palliative and rhetorical legitimating devices.

Socialist systems

There is, as McLaughlin (1993) points out, no simple and universally accepted definition of socialism. A working concept would be social systems in which economic decisions are centrally and bureaucratically organised.

Until the collapse of the Soviet Union in the early 1990s, and the transformation of other former communist states to free market economies at around the same time, it was possible to assert the existence of definite socialist systems. Now the only remaining socialist states are China, North Korea and Cuba, and even these states have moved a long way from the socialist ideals that they once held by embracing many market reforms.

Socialist states have, or have had, characteristic socialist legal systems. These are, or were, based on the Marxist-Leninist doctrine of 'dialectical materialism' that predicted an evolution, by means of proletarian revolution, to a communist state via a socialist intermediary phase. In the final communist utopia it was predicted that the state, and with it law, would wither away (Pashukanis 1978). Incidentally, this view, that a 'world without laws' is possible if human exploitation can be curbed, is still projected by some idealistic political commentators (The World Socialist Party 2000)

According to Marx, the *economic* organisation of production is decisive for the conditions of a society. Law, along with social customs and morality, is just part of the economic *superstructure* (i.e. it is merely an epiphenomenon). From the Marxist point of view, there can be no fair or just laws in capitalist states. Once a just *infrastructure* has been created law will disappear as a useless appendage.

The fact that, in socialist states, law did not wither away has been explained by reference to three factors (David and Brierley 1985). First, retention of law in socialist states was necessary in the interests of national security; that is, protection from enemies real and imagined, external and internal. Encirclement of the socialist states by capitalist countries added impetus to the need for the state to impose its will, coercively, to prevent the backsliding of institutions or individuals to former allegedly exploitative relations.

Second, law was considered necessary to bring about the transformation of the economic system necessary for cessation of intra-human exploitation. This was achieved in the former USSR through the process of total collectivisation of the economy (David and Brierley 1985, 186), including the state control of industry and the radical reorganisation of rural farms into collectives (*kolkhoz*) or large state-run farms (*sovkhozi*).

Third, the minds of people, being allegedly contaminated by capitalist ideologies, had to be kept thinking or made to think along socialist lines. Law, therefore, played an important propaganda or educational role. To this end efforts were made in the USSR to disseminate and popularise the Soviet Constitution, institutions and the law. The Soviet court, likewise, was thought of as a school so that 'a failure has occurred if the condemned party does not approve of his sentence, and if opposing parties do not leave the court reconciled by recognizing the just character of the decision made in application of socialist law' (David and Brierley 1985, 195).

Protection of a shared and state-owned environment was amongst the more important residual functions that law could perform. Indeed, it has been noted by writers such as Goldman (1972) that the environment in socialist states was, at least in the early days of such regimes, protected by a swathe of environmental laws unmatched in capitalist states for many years (Goldman 1972, Appendix A). Pollution standards set in these laws were, on the face of it, very strict indeed. Goldman (1972) comments:

> it may be true, as the Minister of Public Health Petrovsky asserted, that the Soviet Union was the first country in the world to set maximum permissible concentrations for harmful substances in the air. It also was one of the first to set limits on the discharge of various types of water effluent. In both cases the maximum norms were generally lower than those established in other countries in the world.
>
> (Goldman 1972, 24)

The Soviet Union generated an impressive array of public health and conservation laws. Lenin orchestrated the creation of around 15,000 square miles of nature reserves, many of which were strictly protected from human intervention and reserved as baselines for the scientific study of ecosystems (McLaughlin 1993, 50). Many anti-pollution and conservation laws were also passed aiming at high ideals. Article 4 of the Conservation Law of Soviet Republic 1960 (Goldman 1972, Appendix B) gives a taste of their demanding requirements:

- Art. 4 (conservation of waters) 'surface and underground waters are subject to protection against depletion, pollution and obstruction and to regimen regulation as sources of water supply for the population and the national economy . . .

  All organizations whose activity affects the water regimen are required:

  (a) to carry out in the areas used hydromeliorative, forestmeliorate, agrotechnical and sanitary measures that will improve the water regimen and prevent the possibility of harmful effects on water (floods, heating, bogging, salinization, soil erosion, formations of ravines, freshets, etc.);
  (b) to use water resources without exceeding established norms and to expend irrigation, ground and artesian waters prudently, without permitting nonproductive use of them; to avoid the formation of nonproductive shoal water in the construction of reservoirs;
  (c) to build purification installations for artificial or natural purifying at all enterprises that discharge polluted waters into bodies of water;

> (d) to prevent the pollution and silting of spawning grounds and the obstruction of passageways to them as a result of timber floating;
> (e) in designing hydrotechnical projects, to provide for measures that will ensure the protection and reproduction of fish resources.
>
> It is forbidden to put into operation enterprises, shops and installations that discharge sewage without carrying out measures that ensure purifying of it.'

To understand the reasons for the failure of laws such as these to protect the Soviet environment, we have to consider the reasons for the collapse of socialism itself in the Soviet bloc. Sarkar (1999) gives three convincing reasons for the demise of the USSR in 1989. The first was the failure of socialist states to deliver the economic goods that they rashly promised to their populations. The USSR and its satellite states had embarked on a mistaken and impossible mission to match, or even exceed, the economic growth of the rest of the world. The roots of impossibility of this mission lay in the geographical and resource limitations of the country (the USSR, although vast, was fairly unproductive), along with the lack of colonial or neo-colonial (i.e. exploitative) relations with less-developed states. The USSR could never have matched the growth potential of the 'rest of the world'; by failing to do so it created ecological havoc and wholesale disappointment to its inhabitants.

The second factor contributing to the collapse was socialism's failure to create a 'new moral order' in which psychologically superior individuals would seek maximisation of social rather than individual ends. The evidence, if anything, is to the contrary: corruption and nepotism were rife, wealth inequities based on exploitative practices were wider than those in capitalist states, and individuals remained ideologically individualistic. The terroristic rule by the New Class of political elite produced resentment and fear in the general population rather than enhanced ethical precepts.

The third factor was the inescapability of Hardin's 'Tragedy of the Commons Thesis'. As we have already had occasion to note (p. 7), Hardin rightly observes that when things are owned in common, those with rights of access have too much incentive to exploit and too little incentive to conserve, with the end result that everything is spoilt. Ownership of the environment in common is still one of the hallmarks of socialist regimes. For example, Article 26 of the Chinese Constitution still provides:

1 The state protects and improves the living environment and the ecological environment, and prevents and remedies pollution and other public hazards.
2 The state organizes and encourages afforestation and the protection of forests.

And Article 9 of the Constitution declares that

1 Mineral resources, waters, forests, mountains, grassland, unreclaimed land, beaches, and other natural resources are owned by the state, that is, by the whole people, with the exception of the forests, mountains, grassland, unreclaimed land, and beaches that are owned by collectives in accordance with the law.

Unfortunately – and inevitably given the status of commons – the protective attitude towards environmental resources in the USSR was soon lost; the drive for ever-greater economic development resulted in the wholesale nature despoilation, just as Hardin predicted. With the benefit of hindsight it is now clear that such socialist states as have existed have managed their environments very poorly indeed (Sarker 1999, chs 1 and 2; Goldman 1972; Feshbach and Friendly 1992). A few figures indicate the extent of the failure: in 1988 the USSR emitted over twice as much sulphur dioxide per unit GNP than the USA; in 1966 the river Volga was receiving sewage sludge at 300,000 cubic metres per hour; in 1972 Goldman reported that 50 million hectares of Soviet land suffered from erosion, and that mining operations and oil extraction were conducted in a slap-handed and wasteful manner; by the 1970s Siberia had been reduced from a land 'in almost the same condition in which it was created by God' to a place denuded of timber, beset by swamps, river blockages, mud slides, floods, and chronic pollution (Sarker 1999).

In socialist regimes, despite the grand rhetoric, the gap between environmental law and environmental practice grew ever wider as time passed by. In the early days the environment was considered part of the common acquisition of the state on behalf of the whole of the population. In later times, however, environmental regulations were considered merely as examples of what could, *in theory*, be achieved in a socialist regime. There was, in fact, no systematic attempt to enforce such environmental laws as existed: indeed Soviet officials acknowledged that any attempt to do so would defeat the overriding objective of economic growth.

That such socialist systems as have existed have not provided a secure basis for environmental protection does not mean that socialism is totally lost as an engine for environmental sustainability. The body of political theory known as 'ecosocialism' attempts, amongst other things, to spell out the conditions in which left-wing ideas and environmental protection can successfully combine. Ecosocialism takes the view that the solution to environmental problems is to be found in the undoing of the class system and the hierarchical structures that it both creates and is created by. As such, it focuses on aspects of the environment that might appear, from other perspectives, to be less than central to its protection such as urban issues, violence, unemployment and poverty (Pepper 1993). Ecosocialism's focus on care for humanity, and an undoing of exploitation, is by no means linked to an imperative for economic or industrial development which, as we have seen, was amongst the most important of the Soviet bloc's mistakes. It is still possible, therefore, that leftist ideals may yet provide inspiration for legal systems in which environmental protection takes not only theoretical but also a practical priority. As McLaughlin (1993) concludes:

> Socialism theoretically holds great promise for ecologically sound relations between humanity and the rest of nature. The ecological blindness of markets is unnecessary. And since there is no necessity for the future to be discounted, socialistic societies could preserve environments for future generations and adjust to the rhythms of ecological systems. Further, unlike capitalistic economies, socialistic economies have no inherent necessity for economic growth. A socialistic society could decide that growth is a mistake and adopt a steady-state economy.
>
> (McLaughlin 1993, 49)

Eco-anarchist solutions

The third strand in the typology of polities that may provide a basis for robust environmental protection are those characterised as anarchistic or eco-anarchistic. The central tenets of eco-anarchism are:

- a desire to bypass or dismantle the nation-state as the locus of control (i.e. a belief in decentralised governance);
- a claim that eco-anarchist solutions are grounded in ecology; and
- opposition to all types of social domination as well as most human domination of nature (Eckersley 1992).

Bookchin (1982, 1990) – the foremost eco-anarchist – has argued that restoration of proper human–nature relations requires creation of a society in which every individual is capable of participating directly in the formulation of social policy; and that this must be preceded by removal of social hierarchical structures and domination (a view shared by ecosocialists). This, in turn, requires a return to or creation of a society composed of relatively small autonomous units.

Critics of eco-anarchist solutions point to drawbacks with this model. First, according to ecocentric perspectives, it is not correct to identify hierarchical relations as the linchpin of the overall environmental problematique. Rather, one could suggest that reformation of social relations should go hand-in-hand with or perhaps even follow resolution of human–nature relations. Second, even if human domination and nature domination are linked, anarchist solutions do not necessarily provide any good grounds for the resolution of either; feudal and tribal societies – both typically small scale – can be exploitative of both people and nature, and networks of small communities may not provide an effective mechanism to prevent environmental harm. As Eckersley (1992) observes:

> the general ecoanarchist approach of 'leave it all to the locals who are affected' makes sense only when the locals possess an appropriate social and ecological consciousness. It also assumes that bioregion A is not a matter of concern to people in bioregion B . . . historically most progressive social and environmental legislative changes – ranging from affirmative action, human rights protection, and homosexual law reform to the preservation of wilderness areas – have tended to emanate from more cosmopolitan central governments rather than provincial or local decision making bodies.
>
> (Eckersley 1992, 173)

Furthermore, the eco-anarchist web-like, horizontal decision-making structure 'has no *built in* recognition of the "self management" interests of similar or larger social and ecological systems that lie beyond the local community' (1992, 177). As Goodin notes, this means that gains in local decentralised control are offset by increased difficulty in coordination between local communities. For these reasons it is probable that, although ecologically sound in many respects, anarchistic solutions can only be appropriate as a part of a multi-tiered approach in which decisions are adopted at the level most appropriate to their ecological effects.

The connection of law to anarchistic solutions occurs in several ways. If one accepts Bookchin's argument that removal of intra-human domination is a necessary, perhaps sufficient, step to protecting nature, then we can observe a whole raft of common law and legislation that indirectly serves the interests of the environment. Race relations laws, criminalisation of domestic violence, legal protection of fundamental human rights, and so on – all such laws are, on this view, indirectly protecting nature.

As we have already seen, most political/legal systems offer some form of power division between the centre and the periphery. In federal systems this is achieved by reserving powers to local areas and by restricting the legislative competence of central government to certain matters.

Environmental law is often supportive of local activism. As we have seen most legal systems have rules that allow locally based groups rights of standing before the courts. Legal structures that enable local citizens to campaign on local environmental issues and challenge decisions adversely affecting the 'amenity of their area' can be perceived as part of a weak anarchism.

Liberalism

Most environmental law currently exists within what are more or less liberal-democratic societies. These two aspects are, of course, distinguishable: democracy is about registering *public* values in the overall political decision-making process; liberalism is the notion of refraining from the imposition of views of the 'good' life or 'right' behaviour, especially in the so-called 'private sphere' of life. The final section in this book is a consideration of whether these two aspects of modern political structures can accommodate laws that have the strength to alter the current set of globally damaging environmental practices.

A moment's reflection will reveal that liberal-democratic society is not emerging with real solutions to environmental problems. As Plumwood (1995) observes:

> It is a matter of widespread observation that actually-existing liberal-democratic political systems are not responding in more than superficial ways to a state of ecological crisis which everyday grows more severe but which everyday is perceived more as normality.
>
> (Plumwood 1995, 142)

Not all commentors have been as harsh as Plumwood. Mark Sagoff, in particular, has famously defended the compatibility of liberalism with robust environmental protection. According to Sagoff (1995), few environmental laws transgress liberalism's basic demand for a core of private behaviour and beliefs because:

> Environmental decisions, by and large, have to do with what goes on out of doors not indoors; they concern the character and quality of the public household not of the private home . . . Thus the content of environmental policy rarely becomes relevant to the kind of neutrality essential to liberalism.
>
> (Sagoff 1995, 183)

But, in fact, everyday 'private' decisions (e.g. consumer behaviour, lifestyle) are collectively the driving force behind environmental degradation. The 'out of doors' aspects of these private behaviours would not exist otherwise. Take transport for example. Roads damage the environment. Yet roads are only built because of *private* decisions to own and use motor vehicles, and decisions to travel from A to B – often for reasons that are not easy to justify. As Plumwood observes, liberalism

> creates major barriers to corrective ecological action by placing crucial areas of environmental impact beyond the range of democratic correction and reshaping, especially the institutions of accumulation and property.
>
> (Plumwood 1995, 146)

The other side of liberalism that may not be compatible with strong environmentalism is the demand that society should remain neutral between competing conceptions of the good life. It is probably true that a society which is not prepared to use institutions of the state to *advance* a green conception of the good life will be unable to bring about the comprehensive changes in life practices that are required for sustainability. Environmental laws have some function in addressing this point, by standing in the place of ethics and, as such, providing a necessary intermediate phase in the development of a proper social-ecological consciousness.

Democracy

Environmentalists and Greens often assume that democracy (i.e. public participation in the formulation and enforcement of laws and policies) is an unequivocal good. More and better democracy is equated with more and better environmental policies and laws. To an extent this may be true since an enfranchised public is in a better position than a disenfranchised public to exert its environmental values or green ideals. There may be other good grounds for supporting democracy in the context of environmental protection:

- Humans are fallible and get things wrong (Popper 1966). Democracy offers an opportunity for periodic evaluation of government and government policies and, if necessary, replacement of both.
- Democracy may be the best way to achieve major transitions in society peacefully (Hayek 1979).
- Democracy prevents abuse of power and protects the individual's rights and liberties (Hayek 1979; Mill [1868] 1972).
- Democracy, with its constant rotation of government, allows departing administrations to take unpopular steps to protect the environment (as, for example, with President Clinton's eleventh-hour designations of new conservation areas).

There is, nevertheless, a danger in perceiving democracy as a panacea for environmental sickness. In a world of self-interested individuals one may expect jobs and employment – especially at a local level – to be put before environmental protection. Democracy cannot be identified as a green 'good' to be ranked alongside other green values. If and when the public's environmental consciousness is raised to a much higher standard than currently exists it *may* be desirable to increase democratic participation in law and policy processes. Until then, at least as an intermediate stage, less rather than more democracy may be called for. Heilbroner (1975, 110) reluctantly draws the conclusion that 'the passage through the gauntlet ahead may only be possible under governments capable of rallying obedience far more effectively than would be possible in a democratic setting'.

The tendency for democracies to reject robust environmental governance has been illustrated by politics in the United States. Wilson (1999) recounts the way in which the Republican's 'contract with America' during the the 104th Congress resulted in a considerable *retreat* from pre-existing environmental regulations. The general tenor of the Republican programme was removal of red tape and withdrawal of bureaucracy, similar to the UK drive for 'deregulation' in the early 1990s. A significant part of this 'releasing' of businesses was a programme for reduction of environmental legislation. Thus

- New legislation was introduced requiring environment agencies to carry out risk assessment and cost–benefit analysis before introducing new environmental regulations.
- The US EPA's budget was cut.
- The 'Dirty Water bill', as it became known, proposed relaxing federal water pollution standards and removing 80 per cent of wetlands from federal protection.

- A bill proposed review of all national parks and removal of national park status for those 'no longer required'.

(Wilson 1999, 22)

This practical attack was matched by a growth in strong anti-environment language. Congress Representative Don Young is quoted as stating 'I am proud to say that all environmentalists are my enemy . . . they are not Americans' (Wilson 1999, 22). Environmental groups such as the Sierra Club and the League of Conservation Voters responded to this tide of pro-business sentiment by launching their own counter-attacks, including TV and other adverts exposing the anti-environment sentiments of several Republicans. In several cases this contributed to their subsequent electoral defeat.

President Clinton and the Democratic Party learned their lessons from this skirmish by focusing much more sharply on the kinds of environmental legislation they would seek dilution of. The realisation had dawned that in contemporary American society one cannot afford to appear to be either anti-business or anti-environment. Clinton was elected on a manifesto which, perhaps paradoxically, promised both economic growth *and* environmental protection.

This account of the pro- and anti-environmental movements confirms the major weakness of democracy in the environmental context: it is as likely to provide a mechanism for expression of anti-environmental sentiments as for pro-environmental values.

A challenge to the analysis of democracy offered above would be to ask why, if democracy is a route to dominance of business interests, we nevertheless see *so much* environmental law in democratic states? Surely, interest group theory would suggest that, given their superior ability to canvass politicians, the business world would see to it that only a bare minimum of environmental law came to fruition. I have dealt with this issue elsewhere (Wilkinson 1997), but it may be helpful to mention a number of answers to this challenge:

1 Environmentalists show much better ability to campaign as interest groups than standard theory suggests. In particular they are not deterred by the notion of free-riders, and will campaign for benefits for others (whereas businesses tend to behave as self-interested entities) and have achieved good cooperative lobbying strategies.
2 Businesses are not nearly as opposed to environmental legislation as might first be thought: regulation can be a vehicle for self-interest for

businesses, acting as a barrier to other firms and as a mechanism for 'rent seeking'; environmental legislation is no exception. Businesses are fully aware that most environmental law is strong on rhetoric, weak in practice, and will in any event usually be only partially enforced.

3 Businesses are not as able to resist environmental legislation due to limits of organisational competence and conflicts between the interests of corporate and employee goals.

Conclusion

The evolution of better environmental laws may require shifts away from existing political structures and processes. The twentieth-century model of sovereign nation-states, each internally liberal-democratic in orientation, may give way to something as yet undefined. It is probable that if environmental law in the twenty-first century is to be effective, existing political structures will need to be infused with a synergy of ideas from socialist, capitalist and anarchist political thought. Paradoxically, we probably need *both more and less* democracy – increased public opportunities for protection of local environments, but increased powers for regional international legal institutions to legislate to prohibit ecodamage. Similarly, whilst personal freedom to re-engage with nature may need to be increased (as a prerequisite to enhanced ecological sensibility) personal freedoms that depend on nature exploitation may need to be curbed.

Key points

- Capitalist, socialist and anarchist systems each have strengths and weaknesses as contexts in which to situate environmental laws.
- Democracy assists the creation and enforcement of environmental laws by enfranchising public environmental concerns and creating mechanisms of accountability. It suffers, however, from the limitations of popular support for the radical changes in western lifestyle that may be necessary if an adequate response to environmental harm is to occur.
- Liberalism – the backbone of western lifestyle – carves out a private domain where laws may only rarely interpose. Liberal philosophy is desirable, but lifestyle changes previously considered sacrosanct matters of personal taste may be required for strong environmental protection.

Further reading

McHallam, A. (1991) *The New Authoritarians: Reflections on the Greens*, London: London Institute for European Defence and Strategic Studies.

Ophuls, W. and Boyans, W. (1992) *The Politics of Scarcity Revisited*, New York: Freeman.

Westra, L. (1993) 'The Ethics of Environmental Holism and the Democratic State: Are they in Conflict?', *Environmental Values*, vol. 2, 125–36.

The above material provides extension material concerning the important question of whether liberal-democratic systems can provide the required degree of environmental protection.

Discussion questions

1. Consider the epithet 'Think globally, act locally'. Would some form of anarchist political system provide a better context for the legal protection of the environment than the predominantly central government models that currently exist in most states?
2. Is democracy good for laws protective of the environment?
3. Can we rely on technological progress to protect an indefinite continuation of the current western lifestyle, or are more radical changes needed? If so, how can they be obtained?

Glossary

appeal a request to a supervisory court, usually composed of a panel of judges, to overturn the legal ruling of a lower court.

bill a concrete proposal for an item of new legislation.

burden of proof the duty of a party in a lawsuit to persuade the judge or the jury that enough facts exist to prove the allegations of the case. Different levels of proof may be required depending on the type of case.

case law also known as 'common law'. The law created by judges when deciding individual cases.

contract an agreement between two or more parties in which an offer is made and accepted, and each party agrees to give or do something (known as 'consideration'). The agreement need not usually be in writing.

convention a written, legally binding, agreement between states.

damages the financial compensation awarded to someone who suffered an injury or was harmed by someone else's wrongful act.

defendant in criminal cases, the person accused of the crime. In civil matters, the person or organization that is being sued.

directive a form of European Union law in which the EU issues an instruction to its member states to bring about a certain result.

easement gives one party the right to go onto another party's property. Utilities often get easements that allow them to run pipes or phone lines beneath private property.

evidence information presented in testimony or in documents used to persuade the fact finder (the judge or jury) to decide the case for one side or the other.

ex parte an application in a judicial proceeding or public decision made by someone who is not a party to the original proceedings or decision.

joint and several liability liability in which each party out of a number of parties is potentially liable for the full liability of the group as a whole.

jurisdiction (1) the legal authority of a court to hear and decide a case; (2) the geographical or subject area in which a court has power to exercise its authority.

jury persons selected according to law and sworn to inquire into and declare a verdict on matters of fact.

litigation a case, controversy, or lawsuit. Participants in litigation (plaintiffs and defendants) are called litigants.

negligence a failure to use the degree of care that a reasonable person would use under the same circumstances.

opinion a judge's written explanation of a decision of the court or of a majority of judges. Judges' opinions may differ in the same case. An opinion disagreeing with the majority opinion is known as a 'dissenting opinion'. A 'concurring opinion' agrees with the decision of the court.

parties (1) plaintiffs and defendants (petitioners and respondents) to lawsuits, also known as appellants and appellees in appeals; (2) states that, by signature and ratification, become bound by the terms of an international convention.

plaintiff the person who initiates a lawsuit.

precedent a previously decided case that is considered binding in the court where it was issued and in all lower courts in the same jurisdiction.

preponderance of the evidence the level of proof required to prevail in most civil cases. The judge or jury must be persuaded that the facts are more probably one way (the plaintiff's way) than another (the defendant's).

prosecution a criminal law action brought by the state against an individual or corporation.

regulation (1) a legal technique in which an agency is given authority to supervise an industry or sector. (2) A form of law: in the EU a law that is directly applicable in each Member State without further action;

in the UK an item of secondary legislation usually in the form of a Statutory Instrument.

standing the legal right to initiate a lawsuit. To do so, a person must be sufficiently affected by the matter at hand, and there must be a case or controversy that can be resolved by legal action.

statute a law passed by a legislature.

tort a civil wrong that results in an injury to a person or property.

trust property held by one person ('a trustee') to manage for the benefit of another person ('a beneficiary').

verdict the decision of a jury or a judge.

writ a formal written command, issued from the court, requiring the performance of a specific act.

Bibliography

Aalders, M. (1993) 'Regulation and In-company Environmental Management in the Netherlands', *Law and Policy*, vol. 15, pp. 75–94.

Adonis, A. (1991) 'Subsidiarity: Theory of a New Federalism?' in P. King and A. Bosco (eds), *A Constitution for Europe*, London: Lothian Foundation Press.

Adorno, T. and Horkheimer, M. (1972, orig. 1944) *Dialectic of Enlightenment*, New York: Herder and Herder.

Ahuja, A. (2001) 'Live and Extremely Dangerous', *The Times*, 12 March. http://www.thetimes.co.uk/article/0,,74–97331,00.html

Ajijola, A.D. (1989) *Introduction to Islamic law* (3rd edn), Karachi, Pakistan: International Islamic Publishers.

Alder, J. (1993) 'Environmental Impact Assessment – the Inadequacies of English Law', *Journal of Environmental Law*, vol. 5, pp. 203–20.

Alder, J. and Wilkinson, D. (1999) *Environmental Law and Ethics*, London: Macmillan.

Alexander, L. and Kress, K. (1997) 'Against Legal Principles', *Iowa Law Review*, vol. 82, pp. 739–86.

Allen, T. (1993) 'Commonwealth Constitutions and the Right Not to be Deprived of Property', *International and Comparative Law Quarterly*, vol. 42, p. 523.

Allot, P. (1990) *Eunomia*, Oxford: Oxford University Press.

Altham, W.J. (1999) 'Environmental Self-regulation and Sustainable Economic Growth: the Seamless Web Framework', *Eco-Management and Auditing*, vol. 6, no. 2, pp. 61–75.

Anderson, M.R. (1998) 'International Environmental Law in Indian Courts', *Review of European Community and International Environmental Law*, vol. 7, no. 1, pp. 21–30.

Anonymous (1986) 'Is There Life on Mars After All?', *New Scientist*, 31 July, p. 19.

Anonymous (1991) 'Accord on Environment and Development', *Environmental Policy and Law*, vol. 21, pt 5/6, p. 233 and p. 267.

Anonymous (1995) 'Regions, Asia, APEC', *Review of European Community and International Environmental Law*, vol. 4, no. 1, pp. 77–8.

Aquinas, T. (1991) *Summa Theologiae: A Concise Translation* (ed. & trans. Timothy McDermott), London: Methuen.

Arbor, J.L. (1986) 'Animal Chauvinism, Plant-Regarding Ethics and the Torture of Trees', *Australasian Journal of Philosophy*, vol. 64, pp. 335–9.

Arrow, K., Bolin, B., Costanza, R., Dasgupta, P., Folke, C., Holling, C.S., Bengt-Owe, J., Levin, S., Karl-Goran, M., Perrings, C., Pimentel, D. (1995) 'Economic Growth, Carrying Capacity, and the Environment', *Ecological Economics*, vol. 15, pp. 91–5.

Aubert, V. (1966) 'Some Social Functions of Legislation', *Acta Sociologica*, vol. 10, pp. 98–120.

Austin, J. (1954, orig. 1832) *The Province of Jurisprudence Determined*, London: Weidenfeld and Nicolson.

Axelrod, R.S. (1994) 'Subsidiarity and Environmental Policy in the European Community', *International Environmental Affairs*, vol. 6, no. 2, pp. 115–32.

Babcock, H.M. (1995) 'Has the U.S. Supreme Court Finally Drained the Swamp of Takings Jurisprudence? The Impact of *Lucas v. South Carolina Coastal Council* on Wetlands and Coastal Barrier Beaches', *The Harvard Environmental Law Review*, vol. 19, no. 1, pp. 1–68.

Baca, B.J., Lankford, T.E. and Gundlanch, E.R. (1987) 'Recovery of Brittany Coastal Marshes in the Eight Years Following the *Amoco Cadiz* Incident', in *Proceedings*, 1987 Oil Spill Conference, sponsored by the EPA, USCG, and API, American Petroleum Institute Publication 4452, pp. 459–64.

Bailey, P.M. (1999) 'The Creation and Enforcement of Environmental Agreements', *European Environmental Law Review*, June, pp. 170–9.

Baker, J.H. (1990) *An Introduction to English Legal History*, London: Butterworths.

Baker, K.K. (1995) 'Consorting with Forests: Rethinking our Relationship to Natural Resources and How We Should Value their Loss', *Ecology Law Quarterly*, vol. 22, pp. 677–727.

Baldock, D. (1992) 'The Polluter Pays Principle and its relevance to Agricultural Policy in European Countries', *Sociologia Ruralis*, vol. 32, pt 1, pp. 49–65.

Bándi, G. (1996), 'Financial Instruments in Environmental Protection' in G. Winter, *European Environmental Law: A Comparative Perspective*, Aldershot: Dartmouth.

Bar, S. and Albin, S. (2000) 'The "Environmental Guarantee" on the Rise? the amended Article 95 after the revision through the Treaty of Amsterdam', *European Journal of Law Reform*, vol. 2, no. 1, pp. 119–34.

Bardach, E. and Kagan, R.A. (1982) *Going by the Book: The Problem of Regulatory Unreasonableness*, Philadelphia, Pa.: Temple University Press.

Bates, J. (1993) 'Environmental Crime: the Criminal Justice Potential of an Environmental Tribunal' in M. Grant (ed.) *Environmental Litigation: Towards an Environmental Court?*, London: United Kingdom Environmental Lawyers Association.

Baumol, W.J. and Oates, W.E. (1975) *The Theory of Environmental Policy*, Englewood Cliffs, N.J.: Prentice-Hall.

Baxter, W.F. (1974) *People or Penguins: the Case for Optimal Pollution*, New York: Columbia University Press.

Beard, T.R. (1999) *Economics, Entropy and the Environment: the Extraordinary Economics of Nicholas Georgesçu-Roegen*, Cheltenham: Edward Elgar.
Beckerman, W. (1975) 'The Polluter-Pays Principle: Interpretation and Principles of Allocation' in Organisation for Economic Co-operation and Development, *Polluter Pays Principle: Definition, Analysis, Implementation*, Paris: OECD.
Beckerman, W. (1995) *Small is Stupid*, London: Duckworth.
Bell, S. (1997) *Ball and Bell on Environmental Law*, London: Blackstone Press.
Bell, S. and McGillivray, D. (2000) *Environmental Law: The Law and Policy Relating to the Protection of the Environment*, (5th edn) London: Blackstone.
Benfield, M. (1998) 'Ethics and Modern Property Development', paper delivered at the Pacific Rim Real Estate Conference, Perth, Western Australia, January 1998, copy on file with the author.
Benkö, M., de Graaff, W. and Rijnen, G.C.M. (1985) *Space Law in the United Nations*, Dordrecht: Martinus Nijhoff.
Bentham, J. (1891, orig. 1776) *A Fragment on Government*, Oxford: Clarendon Press.
Bentham, J. (1948, orig. 1789) *The Principles of Morals and Legislation*, New York: Hafner.
Bentham, J. (1970) 'Anarchical Fallacies', in A. Meldon (ed.) *Human Rights*, Belmont, Calif.: Wadsworth.
Bern, M. (1990) 'Government Regulation and the Development of Environmental Ethics under the Clean Air Act', *Ecology Law Quarterly*, vol. 17, pp. 539–80.
Biekart, J.W. (1995) 'Environmental Covenants Between Government and Industry: A Dutch NGO's Experience', *Review of European Community and International Environmental Law*, vol. 4, no. 2, pp. 141–9.
Birnie, P.W. and Boyle, A.E. (1992) *International Law and the Environment*, Oxford: Clarendon Press.
Blackstone, W. (Sir) (1768) *Commentaries on the Laws of England*, 4 vols, Oxford: Clarendon Press.
Bloch, J. (1994) 'Conservation in a Concrete Jungle: Political, Legal and Societal Obstacles to Environmental Protection in Hong Kong', *The Georgetown Environmental Law Review*, vol. 6, pp. 593–637.
Blomquist, R.F. (1990a) 'The Logic and Limits of Public Information Mandates under Federal Hazardous Waste Law: A Policy Analysis', *Vermont Law Review*, vol. 14, pp. 558–63.
Blomquist, R.F. (1990b) '"Clean New World: Toward An Intellectual History of American Environmental Law, 1961–1990', 25 *Val. U. L. Review*, vol. 1, no. 2.
Blondel, J. (1973) *Comparative Legislatures*, Englewood Cliffs, N.J.: Prentice-Hall.
Bodansky, D. (1991) 'Scientific Uncertainty and the Precautionary Principle', *Environment*, September, pp. 4–5, 43–4.
Bodansky, D. (1994) 'The Precautionary Principle in US Environmental Law' in T. O'Riordan and J. Cameron (1994) *Interpreting the Precautionary Principle*, London: Earthscan.

Bodansky, D. and Brunnée, J. (1998) 'The Role of National Courts in the Field of International Environmental Courts', *Review of European Community and International Environmental Law*, vol. 7, no. 1, pp. 11–20.

Boehmer-Christiansen, S. (1994) 'The Precautionary Principle in Germany – enabling Government' in T. O'Riordan and J. Cameron, *Interpreting the Precautionary Principle*, London: Earthscan.

Bongaerts, J. (1999), 'Carbon Dioxide Emissions and Cars: An Environmental Agreement at EU Level', *European Environmental Law Review*, April, pp. 101–4.

Bonyhady, T. (1987) *The Law of the Countryside*, Abingdon, UK: Professional Books.

Bonyhady, T. (1992) 'Property Rights' in T. Bonyhady (ed.) *Environmental Protection and Legal Change*, Sydney: Federation Press.

Bookchin, M. (1982) *The Ecology of Freedom*, San Francisco, Calif.: Cheshire Books.

Bookchin, M. (1990) *The Philosophy of Social Ecology: Essays on Dialectical Naturalism*, Montreal: Black Rose Books.

Bosselman, F.P. and Tarlock, A.D. (1994) 'The Influence of Ecological Science in American Law: An Introduction', *Chicago-Kent Law Review*, vol. 69, pp. 847–73.

Botkin, D. (1990) *Discordant Harmonies: A New Ecology for the Twenty-first Century*, Oxford: Oxford University Press.

Boulding, K. (1971) 'The Economics of the Coming Spaceship Earth' in H. Jarrett, *Environmental Quality in a Growing Economy*, Baltimore, Md.: Johns Hopkins.

Bowman, M. (1997) 'New Legislative "Protection" of Voluntary Environmental Audits: Incentive or Indictment?', *Australian Business Law Review*, vol. 27, no. 5, p. 391.

Boyle, A. (1991) 'Saving the World? Implementation and Enforcement of International Environmental Law through International Institutions', *Journal of Environmental Law*, vol. 3, no. 2, pp. 229–45.

Boyle, A. and Anderson, M. (1996) *Human Rights Approaches to Environmental Protection*, Oxford: Clarendon.

Brennan, A. (1995) 'Ecological Theory and Value in Nature' in R. Elliot (ed.) *Environmental Ethics*, Oxford: Oxford University Press.

Brenner, J.F. (1974) 'Nuisance Law and the Industrial Revolution', *Journal of Legal Studies*, vol. 3, pp. 403–33.

Breyer, S. and Heyvaert, V. (2000) 'Institutions for Regulating Risk' in R.L. Revesz, P. Sands and R.B. Stewart, *Environmental Law, the Economy, and Sustainable Development*, Cambridge: Cambridge University Press.

Brinkhorst, L. (1993) 'Subsidiarity and EC Environmental Policy: a Panacea or a Pandora's Box?', *European Environmental Law Review*, vol. 2, no. 17, p. 20.

Brinkman, R., Jasonoff, S. and Ilgen, T. (1985) *Controlling Chemicals*, Ithaca, N.Y.: Cornell University Press.

Brown, V. (1998) 'Environmental Law and Devolution', *Corporate Counsel*, vol. 6, p. 45.

Brubeker, E. (1995) *Property Rights in the Defence of Nature*, London: Earthscan.

Brundtland, G.H. (1987) *Our Common Future: Report of the World Commission on Environment and Development*, Oxford: Oxford University Press.

Brunnée, J. (1998) 'International Environmental Law in Canadian Courts', *Review of European Community and International Environmental Law*, vol. 7, no. 1, pp. 47–56.

Brunton, R. (1995) 'We Must Adopt a Risk-adverse Approach and Always Err on the Side of Caution when Dealing with Environmental Issues' in J. Bennet, *Tall Green Tales*, Perth, Western Australia: Institute of Public Affairs.

Buckland, P.A. (1981) *A History of Northern Ireland*, Dublin: Gill and Macmillan.

Cahen, H. (1988) 'Against the Moral Considerability of Ecosystems', *Environmental Ethics*, vol. 10, no. 3, pp. 195–216.

Calabresi, G. (1991) *A Common Law for the Age of Statutes*, Harvard: Harvard University Press.

Calabresi, G. and Melamed, A.D. (1972) 'Property Rules, Liability Rules and Inalienability. One View of the Cathedral, *Harvard Law Review*, vol. 85, pp. 1089–128

Callicott, J.B. (1980) 'Animal Liberation: A Triangular Affair', *Environmental Ethics*, vol. 2, pp. 311–38.

Callicott, J.B. (1994), 'The Conceptual Foundations of the Land Ethic' in C. Pierce and D. VanDeVeer (eds), *People, Penguins and Plastic Trees* (2nd edn), Belmont, Calif.: Wadsworth.

Cameron, J. (1994) 'The Status of the Precautionary Principle in International Law' in T. O'Riordan and J. Cameron, *Interpreting the Precautionary Principle*, London: Earthscan.

Cameron, J. and Aboucher, J. (1991) 'The Precautionary Principle: A Fundamental Principle of Law and Policy for the Protection of the Global Environment', *Boston College International and Comparative Law Review*, vol. 14, pp. 1–27.

Campbell-Mohn, C., Breen, B. and Futrell, J.W. (1993) *Environmental Law from Resources to Recovery*, St Paul, Minn.: West Publishing.

Carnwarth, R. (1992) 'Environmental Enforcement: the Need for a Specialist Court', *Journal of Planning and Environment Law*, p. 799.

Carnwarth, R. (1999) 'Environmental Litigation – A Way Through the Maze', *Journal of Environmental Law*, vol. 11, no. 1, pp. 3–14.

Carson, R. (1962) *Silent Spring*, Boston: Houghton Mifflin.

Casey-Lefkowitz, S. (1993) 'Non-governmental Organizations (NGOs): Environmental Law Institute', *Yearbook of International Environmental Law*, vol. 4, pp. 604–6.

CEC (Commission of the European Communities) (1980) *The Directive on the Protection of Groundwater Against Pollution Caused by Certain Dangerous Substances 80/68/EEC*, Official Journal of the European Communities, p. 20.

CEC (Commission of the European Communities) (1990) *Commission Report on the Enforcement of Community Environmental Law*, Brussels.

CEC (Commission of the European Communities) (1997) COM(96)500 *Implementing Community Environmental Law: Communication to the Council of the European Union and the European Parliament* (http://www2.unimaas.nl/~egmilieu/docs/96500en.htm).

CEC (Commission of the European Communities) (1999a) *Annual Survey of the Implementation and Enforcement of Community Environmental Law (1996/7)* (http://europa.eu.int/comm/environment/law/as97.htm).

CEC (Commission of the European Communities) (1999b) *COM Documents 1999: 500–513*, Brussels.

Chadwick, E. (1842) 'Report on the Sanitary Condition of the Labouring Population of Great Britain', Report to the Home Secretary by the Poor Law Commissioners, vol. 26, *House of Lords Sessional Papers*.

Cheng, B. (1953) *General Principles of Law as Applied by International Courts and Tribunals*, London: Stevens and Sons.

Churchill, R.R. and Lowe, A.V. (1983) *The Law of the Sea*, Manchester: Manchester University Press.

Cichetti, C.J. and Peck, N. (1989) 'Assessing Natural Resources Damages: the Case Against Contingent Valuation Survey Methods', *Natural Resources and Environment*, vol. 4, p. 6.

Clapp, B.W. (1994) *An Environmental History of Britain*, Harlow, UK: Longman.

Clark, R. (1989) 'State Terrorism: Some Lessons from the Sinking of the "Rainbow Warrior"', *Rutgers Law Journal*, vol. 20, p. 393.

Coase, R. (1960) 'The Problem of Social Cost', *Journal of Law and Economics*, vol. 3, no. 1, pp. 1–44.

Cocks, R. (2000) 'Victorian Foundations?' in J. Lowry and R. Edmunds (eds), *Environmental Protection and the Common Law*, Oxford: Hart Publishing.

Coke, E. (Sir) (1628) *The Institute of the Laws of England, or, a commentary upon Sir Thomas Littleton*, London.

Collier, U. (1997) 'Sustainability, Subsidiarity and Deregulation: New Directions in EU Environmental Policy', *Environmental Politics*, vol. 6, no. 2, pp. 1–21.

Commoner, B. (1971) *The Closing Circle*, New York: Knopf.

Congresslink.org (1999) 'How a Bill becomes Law: the Case of the Civil Rights Act 1964' (http://www.congresslink.org/lessonplans/civrights.html).

Cooper, D. (1992) 'The Idea of Environment' in D.E. Cooper and J.A. Palmer (eds), *The Environment in Question*, London: Routledge.

Cornish, W.R. and Clark, G.N. (1989) *Law and Society in England, 1750–1950*, London: Sweet and Maxwell.

Coulson, N.J. (1978) *A History of Islamic Law*, Edinburgh: Edinburgh University Press.

Council of the European Communities (1975) 'Council Recommendation of 3 March 1975 Regarding Cost Allocation and Action by Public Authorities on

Environmental Matters', 18 *Official Journal of the European Communities*, vol. L 194, p. 1.

Cranor, C.F., Fischer, J.G. and Eastmond, D.A. (1996), 'Judicial Boundary Drawing and the Need for Context-Sensitive Science in Toxic Torts after Daubert v. Merrell Dow Pharmaceuticals, Inc., *Virginian Environmental Law Journal*, vol. 16, pt 1, pp. 1–77.

Cross, F.B. (1996) 'Paradoxical Perils of the Precautionary Principle', *Washington and Lee Law Review*, vol. 53, no. 3, pp. 851–925.

D'Amato, A. (1995) *International Law: Process and Prospect* (2nd edn), New York: Transnational Publishers.

D'Amato, A. and Chopra, S.K. (1991) 'Whales: Their Emerging Right to Life', *American Journal of International Law*, vol. 85, pp. 21–62.

Dalton, D.S. (1993) 'The Negotiated Rulemaking Process – Creating a New Legitimacy in Regulation', *Review of European Community and International Environmental Law*, vol. 2, no. 4, pp. 354–61.

Daly, H.E. (1992) *Steady-State Economics* (2nd edn), London: Earthscan.

Dashwood, A. (1998) 'States in the European Union', *European Law Review*, vol. 23, pp. 201–16.

David, R. and Brierley, J.E.C. (1985) *Major Legal Systems in the World Today*, (3rd edn), London: Stevens and Sons.

Davidson, S. (1991) 'The *Rainbow Warrior* Arbitration Concerning the Treatment of French Agents Mafart and Prieur', *International and Comparative Law Quarterly*, vol. 40, p. 446.

Dawkins, R. (1989) *The Selfish Gene*, Oxford: Oxford University Press.

De Cruz, P. (1995) *Comparative Law in a Changing World*, London: Cavendish.

Delgardo, R. (1982) 'Beyond Sindell: Relaxation of Cause-in-Fact Rules for Indeterminate Plaintiffs', *California Law Review*, vol. 70, pp. 881–908.

Department of the Environment (1990) *This Common Inheritance*, London: HMSO.

Desgagné, R. (1995) 'Integrating Environmental Values into the European Convention on Human Rights', *American Journal of International Law*, vol. 89, pp. 263–94.

Des Jardins, J.R. (1997) *Environmental Ethics: An Introduction to Environmental Philosophy* (2nd edn), Belmont, Calif.: Wadsworth.

De-Shalit, A. (1997) *Where Philosophy Meets Politics: The Concept of the Environment*, Oxford Centre for the Environment, Ethics and Society, Research Paper No. 13, Oxford: OCEES.

De Smith, S.A. (1981) *Constitutional and Administrative Law*, London: Penguin Books.

Devall, B. and Sessions, G. (1984) 'The Development of Natural Resources and the Integrity of Nature', *Environmental Ethics*, vol. 6, pp. 296–322.

Diamond, B.K. (1968) 'The Fallacy of the Impartial Expert' in R.C. Allen (ed.) *Readings in Law and Psychiatry*, Baltimore, Md.: Johns Hopkins.

Dias, A. (1994) 'Judicial Activism in the Development and Enforcement of Environmental Law', *Journal of Environmental Law*, vol. 6, no. 2, pp. 243–62.

Dicey, A.V. (1965) *Introduction to the Study of the Law of the Constitution* (10th edn), London: Macmillan.

Dimento, J. (1986) *Environmental Law and American Business: Dilemmas of Compliance*, New York: Plenum Press.

Dimento, J. (1989) 'Can Social Science Explain Organizational Non-compliance with Environmental Law?', *Journal of Social Issues*, vol. 45, pp. 109–32.

Dixon, M. (1990) *Textbook on International Law*, London: Blackstone Press.

Dobson, A. (1990) *Green Political Thought*, London: Routledge.

Dodson, M. (1994) 'New Perspectives for Aboriginal Land Rights' in L. Van der Vlist (ed.) *Voices of the Earth: Indigenous Peoples, New Partners and the Right to Self-determination in Practice*, Amsterdam: Netherlands Centre for Indigenous Peoples.

Doherty, M. (1999) 'The Status of the Principles of EC Environmental Policy', *Journal of Environmental Law*, vol. 11, no. 2, pp. 354 –86.

Doherty, M. (2000) 'The Judicial Use of the Principles of EC Environmental Policy', *Environmental Law Review*, vol. 2, no. 4, pp. 251–63.

Douma, W.T. (1996) 'The Precautionary Principle', *Úlfljótur*, vol. 49, nos 3/4, pp. 417–30.

Douma, W.T. (1999) 'The Beef Hormones Dispute and the Use of National Standards under WTO Law', *European Environmental Law Review*, May, pp. 137–44.

Dower, N. (1994) 'The Idea of the Environment' in R. Attfield and A. Belsey, *Philosophy and the Natural Environment*, Cambridge: Cambridge University Press.

Driessen, B. (1999) 'New Opportunities or Trade Barrier in Disguise? The EC Eco-labelling Scheme', *European Environmental Law Review*, January, pp. 5–15.

Dudek, D.J. and Palmisano, J. (1988) 'Emissions Trading: Why is this Thoroughbred Hobbled?', *Columbia Journal of Environmental Law*, vol. 13, pp. 217–56.

Dupuy, P. (1991) 'Soft Law and the International Law of the Environment', *Michigan Journal of International Law*, vol. 12, pp. 420–35.

Dutton, Y. (1996) 'Islam and the Environment: a Framework for Enquiry' in C. Lamb (ed.) *Faiths in Dialogue: Number One, Faiths and the Environment*, London: Centre for Inter-Faith Dialogue.

Dworkin, R. (1967) 'Is Law a System of Rules?', *University of Chicago Law Review*, vol. 14, pp. 22–9.

Dworkin, R. (1977) *Taking Rights Seriously*, London: Duckworth.

Dworkin, R. (1984) 'Rights as Trumps' in J. Waldron (ed.) *Theories of Rights*, New York: Oxford University Press.

Eckersley, R. (1992) *Environmentalism and Political Theory: Toward an Ecocentric Approach*, London: UCL Press.

Emilou, N. (1992) 'Subsidiarity: An Effective Barrier Against the Enterprises of Ambition?', *European Law Review*, vol. 27, no. 5, pp. 383–407.

Emmenegger, S. and Tschentscher, A. (1994) ‘Taking Nature’s Rights Seriously: the Long Way to Biocentrism in Environmental Law’, *Georgetown International Environmental Law Review*, vol. 6, pp. 545–92.
Emond, D.P. (1984) ‘Co-operation in Nature: A New Foundation for Environmental Law’, *Osgoode Hall Law Journal*, vol. 22, no. 2, pp. 323–48.
ENDS (1992) *Dangerous Substances in Water: a Practical Guide*, London: Environmental News Data Services.
ENDS (1993) ‘Improving the Chemical Industry’s Performance’, *ENDS Report*, no. 223, pp. 16–19.
ENDS (1996) ‘Lords Call for Easier Access to Environmental Information’, *ENDS Report*, no. 262, pp. 27–9.
ENDS (1998a) ‘Landmark IPC Prosecution Rocks Agency Enforcement Policy’, *ENDS Report*, no. 286, pp. 17–19.
ENDS (1998b) ‘ISO 14001 Certification Tops 700 as EMAS Trails Behind’, *ENDS Report*, no. 286, p. 8.
ENDS (1999a) ‘Environment Struggles for a Place in Devolved Assemblies’, *ENDS Report*, no. 296, pp. 35–6.
ENDS (1999b) ‘Review of Tanker Salvage Stirs Controversy over Prosecution Policy’, *ENDS Report*, no. 290, pp. 41–2.
ENDS (1999c) ‘Agency Plans Shift to Standardise Permits’, *ENDS Report*, no. 297, p. 405.
ENDS (1999d) ‘Setback for EMAS as Momentum Builds Behind ISO14001’, *ENDS Report*, no. 291, p. 11.
ENDS (1999e) ‘Ford, GM Push ISO 14001 Down their Supply Chains’, *ENDS Report*, no. 297, p. 39.
ENDS (1999f) ‘A Big Step Forward for Green Tax Reform’, *ENDS Report*, no. 290, pp. 17–22.
ENDS (1999g) ‘Water Pollution Charges Dropped but Pesticides Tax Draws Closer’, *ENDS Report*, no. 290, p. 21.
ENDS (1999h) ‘Budget Package Tackles Transport Impacts’, *ENDS Report*, no. 290, p. 25.
ENDS (1999i) ‘Energy Tax, Emissions Trading Debate Hots Up’, *ENDS Report*, no. 292, pp. 21–3.
ENDS (2000a) ‘IPPC “General Binding Rules” Takes Shape’, *ENDS Report*, no. 305, p. 43.
ENDS (2000b), ‘Time for Mandatory Environmental Reporting, Committee Told’, *ENDS Report*, no. 305, pp. 33–4.
Environment Agency (1998) ‘Enforcement and Prosecution Policy’ (http://www.environment-agency.gov.uk).
Environment Agency (1999) *Guidance on the Application of Environmental Risk Assessment for Waste Management Licensing: Guidance Note no. 25*, London: HMSO.
Environmental Protection Agency (http://www.epa.gov/ceisweb1/ceishome/ceisdocs/tri/tri-exec.htm)

Epstein, R.A. (1979) 'Possession as the Root of Title', *Georgia Law Review*, vol. 13, pp. 1221–43.

Ehrlich, E. (1936) *Fundamental Principles of the Sociology of Law*, trans. W.L. Moll, Cambridge, Mass.: Harvard University Press.

EU Parliament, COM Docs (http://www.europarl.eu.int/scripts/dg7/ticom.exe/?t=i&l=en&s=00&p=dg7/ticom/data/gpar.txt&v=ft).

Evans, D. (1992) *A History of Nature Conservation in Britain*, London: Routledge.

Farber, D.A. (1992) 'Politics and Procedure in Environmental Law', *Journal of Law, Economics and Organisation*, vol. 8, no. 1, pp. 61–81.

Farber, D.A. (1999) 'Taking Slippage Seriously: Noncompliance and Creative Compliance in Environmental Law', *The Harvard Environmental Law Journal*, vol. 23, no. 2, pp. 297–326.

Faure, M. (1998) 'Harmonisation of Environmental Law and Market Integration: Harmonising for the Wrong Reasons?', *European Environmental Law Review*, vol. 7, no. 6, pp. 169–75.

Favre, D. (1993) 'Debate within the CITES Community: What Direction for the Future?', *Natural Resources Journal*, vol. 33, no. 4 (SI), pp. 875–918.

Fawcett, J.E.S. (1968) *International Law and the Uses of Outer Space*, Manchester: Manchester University Press.

Feddersen, C.T. (1999) 'Recent EC Environmental Legislation and its Compatibility with WTO Rules: Free Trade or Animal Welfare Trade?', *European Environmental Law Review*, July, pp. 207–15.

Feinberg, J. (1974) 'The Rights of Animals and Unborn Generations' in William Blackstone (ed.), *Philosophy and the Environmental Crisis*, Athens, Ga.: University of Georgia Press.

Feshbach, M. and Friendly, A. (1992) *Ecocide in the USSR: Health and Nature Under Siege*, New York: Basic Books.

Findley, R.W. and Farber, D.A. (1992) *Environmental Law in a Nutshell*, St Paul, Minn.: West Publishing.

Finnis, J. (1980) *Natural Law and Natural Rights*, Oxford: Clarendon.

Fisk, D. (1998) 'Environmental Science and Environmental Law', *Journal of Environmental Law*, vol. 10, no. 1, pp. 3–8.

Fortuna, R. and Lennett, D. (1987) *Hazardous Waste Regulation: the New Era*, New York: McGraw-Hill.

Foster, N. (1993) *German Law and Legal System*, London: Blackstone Press.

Francis, L.P. and Norman, R. (1978) 'Some Animals are more Equal than Others', *Philosophy*, vol. 53, pp. 507–27.

Frangos, J. (1999) 'Environmental Science and the Law', *Environmental and Planning Law Journal*, vol. 16, no. 2, pp. 175–81.

Frank, J. (1970) *Law and the Modern Mind*, Gloucester, Mass.: Peter Smith.

Frankena, W.K. (1979) 'Ethics and the Environment' in K.E. Goodpaster and K.M. Sayre, *Ethics and Problems in the 21st Century*, Notre Dame, Ind.: University of Notre Dame Press.

Freeman, G.C. (1986) 'Inappropriate and Unconstitutional Retroactive Application of Superfund Liability', *Business Lawyer*, vol. 42, pp. 215–48.

Frey, R.G. (1980) *Interests and Rights: The Case Against Animals*, Oxford: Clarendon Press.

Freyfogle, E.T. (1993) 'Ethics, Community, and Private Land', *Ecology Law Quarterly*, vol. 23, pp. 632–61.

Freyfogle, E.T. (1994) 'The Ethical Strands of Environmental Law', *University of Illinois Law Review*, 4, pp. 819–846.

Friendly, H.J. (1977) 'Federalism: A Foreword', *Harvard Environmental Law Review*, vol. 88, p. 1019.

Friends of the Earth (FoE) UK (2000) 'Factory Watch' (http://www.foe.co.uk/campaigns/industry_and_pollution/factorywatch/index.html).

Futrell, J.W. (1993) 'The History of Environmental Law' in C. Campbell-Mohn, B. Breen and J.W. Futrell, *Environmental Law from Resources to Recovery*, St Paul, Minn.: West Publishing.

Gaines, S.E. (1991) 'The Polluter-Pays Principle: From Economic Equity to Environmental Ethos', *Texas International Law Journal*, vol. 26, no. 3, pp. 463–96.

Georgesçu-Roegen, N. (1971) *The Entropy Law and the Economic Process*, Cambridge, Mass.: Harvard University Press.

Gergen, M. (1994) 'The Failed Promise of the Polluter Pays Principle: an Economic Analysis of Landowner Liability for Hazardous Waste', *New York University Law Review*, vol. 69, pt 3, pp. 624–91.

Gewirth, A. (1981) 'The Basis and Content of Human Rights' in J.R. Pennock and J.W. Chapman (eds) *Human Rights: Nomos XXIIII*, New York: New York University Press.

Gillette, C. and Krier, J. (1990),'Risk, Courts and Agencies', *University of Pennsylvania Law Review*, vol. 138, pp. 1027–79.

Gilligan, C. (1982) *In a Different Voice*, Cambridge, Mass.: Harvard University Press.

Ginsberg, B.S. and Cummis, C. (1996) 'EPA's Project XL: A Paradigm for Promising Regulatory Reform', *Environmental Law Reporter*, vol. 26, p. 10,059.

Goldfarb, W. (1988) *Water Law*, Chelsea, Mich.: Lewis Publishers, Inc.

Goldman, M.I. (1972) *The Spoils of Progress: Environmental Pollution in the Soviet Union*, Chelsea, Mich.: Lewis Publishers, Inc.

Goldsmith, E. (1972) *A Blueprint for Survival*, Harmondsworth: Penguin.

Goldsmith. E. (1993) *The Way: An Ecological World-view*, Boston, Mass.: Shambhala Publications.

Goldstein, A.S. and Marcus, M. (1977) 'The Myth of Judicial Supervision in Three "Inquisitorial" Systems: France, Italy and Germany', *The Yale Law Journal*, vol. 87, no. 2, pp. 240–83.

Golub, J. (1996) 'Sovereignty and Subsidiarity in EU Environmental Policy', *Political Studies*, vol. 44, no. 4, p. 686.

Goodpaster, K. (1978) ‘On Being Morally Considerable’, *Journal of Philosophy*, vol. 75, pp. 308–25.
Gouldson, A. and Murphy, J. (1998) *Regulatory Realities: the Implementation and Impact of Industrial Environmental Regulation*, London: Earthscan.
Granqvist (1996) ‘The Limits of Space Law’ (http://www.users.wineasy.se/dg/spacelaw.htm#n3).
Grant, M. (1996) ‘Environmental Liability’ in G. Winter, *European Environmental Law: A Comparative Perspective*, Aldershot: Dartmouth.
Grant, M. (1999) ‘Environment Court Project, Final Report’ (http://www.planning.detr.gov.uk/court/index.htm).
Gray, J.S. (1990) ‘Statistics and the Precautionary Principle’, *Marine Pollution Bulletin*, vol. 21, p. 174.
Gray, K. (1991) ‘Property in Thin Air’, *Cambridge Law Journal*, vol. 50, no. 2, pp. 253–307.
Gray, K. (1994) ‘Equitable Property’, *Current Legal Problems*, vol. 47, pt 2, pp. 157–214.
Gray, K. (1999) *Land Law*, London: Butterworths.
Greenberg, M.R. and Schneider, D.F. (1995) ‘Gender Differences in Risk Perception: Effects Differ in Stressed vs Non-stressed Environments’, *Risk Analysis*, vol. 15, no. 4, p. 503.
Gresser, J. (1981) *Environmental Law in Japan*, Cambridge, Mass.: MIT Press.
Greve, M.S. (1989) ‘The Non-Reformation of Administrative Law – Standing to Sue and Public Interest Litigation in West German Environmental Law’, *Cornell International Law Journal*, vol. 22, pp. 197–244.
Grierson, K.W. (1992) ‘The Concept of Species and the Endangered Species Act’, *Virginia Environmental Law Review*, vol. 11, pp. 463–98.
Grubb, M. *et al.* (1993) *The ‘Earth Summit’ Agreements: A Guide and Assessment*, London: Earthscan. (http://www.unep.ch/iucc/conv-e.html), (http://sedac.ciesin.org/pidb/texts/biodiversity.1992.html), (http://sedac.ciesin.org/pidb/texts/rio.declaration.1992.html).
Gruen, L. (1991) ‘Animals’ in P. Singer (ed.) *A Companion to Ethics*, Oxford: Blackwell.
Guha, R. (1989) ‘Radical American Environmentalism and Wilderness Preservation: A Third World Critique’, *Environmental Ethics*, vol. 11, no. 1, pp. 71–83.
Gullet, W. (1997) ‘Environmental Protection and the Precautionary Principle’, *Environmental and Planning Law Journal*, vol. 14, pt 1, pp. 52–69.
Gundling, L. (1990) ‘The Status in International Law of the Principle of Precautionary Action’, *International Journal of Estuarine and Coastal Law*, vol. 5, p. 23.
Gunn, A.S. (1980) ‘Why Should We Care About Rare Species?’, *Environmental Ethics*, vol. 2, pp. 17–37.
Gunningham, N. (1995) ‘Environment, Self-regulation, and the Chemical Industry: Assessing Responsible Care’, *Law and Policy*, vol. 17, no. 1, pp. 57–109.

Guruswamy, L., Palmer, W.R. and Weston, B.H, (1994) *International Environmental Law and World Order*, St. Paul, Minn.: West.
Haas, P.M. (1989) 'Do Regimes Matter? Epistemic Communities and Mediterranean Pollution Control', *International Organization*, vol. 45, pp. 152–76.
Hahn, R.W. and Hester, G.L. (1989) 'Where did all the Markets Go?: an Analysis of EPA's Emissions Trading Program', *Yale Journal on Regulation*, vol. 6, pp. 109–53.
Haigh, N. (1994) 'The Environment as a Test Case for Subsidiarity', *Environmental Liability*, vol. 2, no. 2, pp. 22–5.
Haley, A.G. (1963*) Space Law and Government*, New York: Appleton-Century-Crofts.
Handl, G. (1990) 'Environmental Security and Global Change: the Challenge to International Law', *Yearbook of International Environmental Law*, vol. 3, p. 22.
Hanley, N. and Milne, J. (1996) 'Ethical Beliefs and Behaviour in Contingent Valuation Surveys', *Journal of Environmental Planning and Management*, vol. 39, pp. 255–72.
Hannigan, J.A. (1995) *Environmental Sociology: A Social Constructionist Perspective*, London: Routledge.
Hansard Society, The (1992) *Making the Law – the Report of the Hansard Commission on the Legislative Process*, London: The Hansard Society.
Hardin, G. (1968) 'The Tragedy of the Commons', *Science*, vol. 162, pp. 1243–8, reprinted in A. Markandya and J. Richardson (eds) (1992) *The Earthscan Reader in Environmental Economics*, London: Earthscan.
Hardin, G. (1977) *The Limits of Altruism*, Bloomington, Ind.: Indiana University Press.
Hargrove, E.C. (1989) *Foundations of Environmental Ethics*, Englewood Cliffs, N.J.: Prentice-Hall.
Harris, P. (1993) *An Introduction to Law*, London: Weidenfeld and Nicolson.
Harris, R.H. (1996) 'Promoting Self-Compliance: An Examination of the Debate over Legal Protection for Environmental Audits', *Ecology Law Review*, vol. 23, pp. 663–721.
Hart, H.L.A. (1961) *The Concept of Law*, Oxford: Clarendon Press.
Harvey, L.D. (2000) *Global Warming: The Hard Science*, Harlow, UK: Prentice-Hall.
Hatch Hodge, S. (1997) 'Satellite Data and Environmental Law: Technology Ripe for Litigation Application', *Pace Environmental Law Review*, vol. 14, no. 2, pp. 691–731.
Hawkins, K. (1984) *Environment and Enforcement: Regulation and the Social Definition of Pollution*, Oxford: Clarendon Press.
Hayek, F.A. von (1979) *Law, Legislation and Liberty*, Chicago: University of Chicago Press.
Hays, S. (1969) *Conservation and the Gospel of Efficiency: The Progressive Conservation Movement*, Cambridge, Mass.: Harvard University Press.

Heesterman, W. (1993) *Using the LEXIS full text legal retrieval system*, Warwick: CTI Law Technology Centre.
Heffernan, J. (1982) 'The Land Ethic: A Critical Appraisal', *Environmental Ethics*, vol. 4, pp. 235–48.
Heidt, R. (1990) 'Corrective Justice from Aristotle to Second Order Liability: Who Should Pay When the Culpable Cannot?', *Washington and Lee Law Review*, vol. 47, pp. 347–77.
Heilbroner, R.L. (1975) *An Inquiry into the Human Prospect*, London: Calder & B.
Helm, D. and Pearce, D. (1990) 'Assessment: Economic Policy Towards the Environment', *Oxford Review of Economic Policy*, vol. 6, no. 1, pp. 1–16.
Hern, J. (1992) *The Independent*, 30 March.
Herrup, A. (1999) 'Eco-labels: Benefits Uncertain, Impacts Unclear?', *European Environmental Law Review*, May, p. 144.
Hey, E. (1992) 'The Precautionary Concept in Environmental Law and Policy: Institutionalizing Caution', *The Georgetown International Environmental Law Review*, vol. 4, pp. 303–18.
Hey, E. (1998) 'The European Community's Courts and International Environmental Agreements', *Review of European Community and International Environmental Law*, vol. 7, no. 1, pp. 4–10.
Hilson, C. (1993) 'Discretion to Prosecute and Judicial Review', *The Criminal Law Review*, pp. 739–47.
HMSO (1975) *Report of the Committee on the Preparation of Legislation, Cmnd 6053*, London: HMSO.
HMSO (1999a) 'The Parliamentary Stages of a Government Bill' (http://www.parliament.uk/commons/lib/fs01.pdf).
HMSO (1999b) 'Private Members' Bill Procedure' (http://www.parliament.uk/commons/lib/fs04.pdf).
HMSO (1999c) 'Public Bills before Parliament' (http://www.parliament.the-stationery-office.co.uk/pa/pabills.htm).
Hobbes, T. (1996, orig. 1651) *Leviathan*, Oxford: Oxford University Press.
Hodge, S.H. (1997) 'Satellite Data and Environmental Law: Technology Ripe for Litigation Application', *Pace Environmental Law Review*, vol. 14, no. 2, p. 691.
Hohfeld, W.N. (1919) *Fundamental Legal Conceptions as Applied in Judicial Reasoning*, New Haven, Conn.: Yale University Press.
Holder, J. (1991) 'Regulating Green Advertising in the Motor Car Industry', *Journal of Law and Society*, vol. 18, no. 3, pp. 323–46.
Howarth, W. (1997) 'Self-Monitoring, Self-Policing, Self-Incrimination and Pollution Law', *Modern Law Review*, vol. 60, pp. 200–29.
Huber, P. (1985) 'Safety and the Second Best: The Hazards of Public Risk Assessment in the Courts', *Columbia Law Review*, vol. 85, p. 277.
Huffmann, J.L. (1992) 'Do Species and Nature Have Rights?', *Public Land Law Review*, vol. 13, pp. 51–76.

Hughes, D. (1991) 'The NRA Scheme of Charges in Respect of Discharges into Controlled Waters', *Land Management and Environmental Law Report*, vol. 3, no. 4, pp. 115–17.

Hughes, D. (1995) 'The Status of the Precautionary Principle in Law', *Journal of Environmental Law*, vol. 7, no. 2.

Hughes, L. (1999) 'Environmental Impact Assessment in the Environment Protection and Biodiversity Conservation Act 1999', *Environmental and Planning Law Journal*, vol. 16, no. 5, p. 441.

Hunter, R., Hendrickx, F. and Muylle, M. (1998) 'Environmental Enforcement in Europe', *European Environmental Law Review*, February, pp. 47–55.

Hutchinson, A.C. and Monahan, P. (eds) (1987) *The Rule of Law: Ideal or Ideology*, Toronto: Carswell.

Iles, W. (1991) 'Legislative Drafting Practices in New Zealand', *Statute Law Review*, vol. 12, no. 1.

Iles, W. (1992) 'New Zealand Experience of Parliamentary Scrutiny of Legislation', *Statute Law Review*, vol. 12, no. 3.

Jackson, S.M. (1998) 'Technical Advisors Deserve Equal Billing with Court Appointed Experts in Novel and Complex Scientific Cases: Does the Federal Judicial Center Agree?', *Environmental Law*, vol. 28, pt 2, pp. 431–65.

Jamieson, D. (1984) 'The City Around Us' in T. Regan, *Earthbound: Introductory Essays in Environmental Ethics*, Prospect Heights, Ill.: Waveland Press.

Jans, J. (1996) 'Objectives and Principles of EC Environmental Law' in G. Winter, *European Environmental Law: A Comparative Perspective*, Dartmouth: Aldershot.

Jarvis, J. and Fordham, M. (1993) *Lender Liability*, London: Cameron May.

Jasanoff, S. (1991) 'Cross-National Differences in Policy Implementation', *Evaluation Review*, vol. 15, pp. 103–19.

Jasper, W.F. (1996) *Global Tyranny . . . Step By Step*, Chapter 5, 'The Drive for World Government' (http://www.freedomdomain.com/neworder/tyranny01.html).

Jaspers, K. ([1933] 1951) *Man in the Modern Age* (trans. E. and C. Paul), London: Routledge and Kegan Paul.

Jenkins, L. (1993) 'Trade Sanctions: An Effective Enforcement Tool', *Review of European Community and International Environmental Law*, vol. 2, no. 4, pp. 362–9.

Jewell, T. (1999) 'Setting Environmental Standards', *Environmental Law and Management*, vol. 11, no.1, pp. 31–4.

Johns, C.H.W. (1999) 'Babylonian Law – The Code of Hammurabi' (www.yale.edu/lawweb/avalon/hammpre.htm).

Johnson, P.M. and Beaulieu, A. (1996) *The Environment and NAFTA: Understanding and Implementing the New Continental Law*, Washington, DC: Island Press.

Kassov, O. (1993) 'Environmental Law in Russia', *Review of European Community and International Environmental Law*, vol. 3, pp. 21–2.

Keng, E. (1997) 'Population Control Through the One-Child Policy in China: Its Effects on Women', *Women's Rights Law Reporter*, vol. 18, no. 2, p. 205.

Kheel, M. (1985) 'The Liberation of Nature: A Circular Affair', *Environmental Ethics*, vol. 7, pp. 135–49.

Kibel, P. S. (1999) *The Earth on Trial: Environmental Law on the International Stage*, Routledge: New York.

Kimball, L. (1993) 'Environmental Law and Policy in Antarctica' in P. Sands (ed.) *Greening International Law*, London: Earthscan.

Kirkpatrick, D. (1997) 'Land Use and Sub-division' in D.A.R. Williams, *Environmental and Resource Management Law in New Zealand*, Wellington: Butterworths.

Knight, F. (1921) *Risk, Uncertainty and Profit*, Boston, Mass.: Houghton Mifflin.

Konar, S. and Cohen, M.A. (1997) 'Information as Regulation: The Effect of Community Right to Know Laws on Toxic Emissions', *Journal of Environmental Economics and Management*, vol. 32, no. 1, p. 109.

Koppen, I. (1994) 'Ecological Covenants: Regulatory Informality in Dutch Waste Reduction Policy' in G. Teubner, L. Farmer & D. Murphy (eds) *Environmental Law and Ecological Responsibility*, London: Wiley.

Krämer, L. (1993) *European Environmental Law Casebook*, London: Sweet and Maxwell.

Krämer, L. (1995) *E.C. Treaty and Environmental Law*, London: Sweet and Maxwell.

Krämer, L. (1996) 'Public Interest Litigation in Environmental Matters Before European Courts', *Journal of Environmental Law*, vol. 8, no. 1, pp. 1–18.

Krämer, L. (2000) *E.C. Environmental Law*, (4th edn), London: Sweet and Maxwell.

Kubasek, N. (1998) 'Mandatory Environmental Auditing: A Better Way to Secure Environmental Protection in the United States and Canada', *Journal of Land, Resources, and Environmental Law*, vol. 18, no. 2, p. 261.

Kubasek, N.K. and Silverman, G.S. (2000) *Environmental Law*, Englewood Cliffs, N.J.: Prentice-Hall.

Landes, W.M. and Posner, R.A. (1987) *The Economic Structure of Tort Law*, Cambridge, Mass.: Harvard University Press.

Lang, J.C. (1999) 'Legislative, Regulatory and Juridical Dilemmas in Environmental Auditing', *Eco Management and Auditing*, vol. 6, no. 3, pp. 101–14.

Lau, M. (1995) 'The Scope and the Limits of Environmental Law in India', *Review of European Community and International Environmental Law*, vol. 4, no. 1, pp. 15–21.

Lavrysen, L. (1998) 'The Precautionary Principle in Belgian Jurisprudence: Unknown, Unloved?', *European Environmental Law Review*, March, pp. 75–82.

Lavrysen, L. (1999) 'The Codification of Flemish Environmental Law', *European Environmental Law Review*, vol. 8, nos. 8/9, pp. 230–3.

Legal Aid Board (2000) 'The Funding Code' (www.legalservices.gov.uk/stat/fc3_cr2.pdf).

Legge, D. (1992) 'The Social Impact of Water Metering', *Utilities Law Review*, vol. 3, no. 4, pp. 152–5.

Leggett, J. (1999) *The Carbon War: Global Warming and the End of the Oil Era*, London: Penguin.

Leopold, A. (1949) *A Sand County Almanac; and Sketches Here and There*, New York: Oxford University Press.

LEXIS (1999) 'LEXIS Research System' (http://www.lexis.com).

Lis, J. and Chilton, K. (1993) 'Limits of Pollution Prevention', *Society*, Mar/Apr, at 49.

Little, G. (2000) 'Scottish Devolution and Environmental Law', *Journal of Environmental Law*, vol. 12, no. 2, pp. 155–74.

Locke, J. (1988, orig. 1690) *Two Treatises of Government*, P. Laslett (ed.), Cambridge: Cambridge University Press.

Lovelock, J.E. (1979) *Gaia: A New Look at Life on Earth*, Oxford: Oxford University Press.

Lunar Corp (1999) 'Prospecting for Water on the Moon' (http://www.lunacorp.com/water.html).

Lyons, D.L. (1984) 'Utility and Rights' in J. Waldron (ed.) *Theories of Rights*, New York: Oxford University Press.

Lyons, J.J. (1986) 'Deep Pockets and CERCLA: Should Superfund Liability be Abolished?', *Stanford Environmental Law Journal*, vol. 6, pp. 271–344.

Lyster, S. (1985) *International Wildlife Law*, Cambridge: Grotius.

McAuslan, P. (1991) 'The Role of Courts and Other Judicial Type Bodies in Environmental Management', *Journal of Environmental Law*, vol. 3, no. 2, pp. 195–207.

McCloskey, H.J. (1979) 'Moral Rights and Animals', *Inquiry*, vol. 22, pp. 23–54.

McCormick, J. and McDowell, E. (1999) 'Environmental Beliefs and Behaviour in Scotland', in E. McDowell and J. McCormick, *Environment and Scotland: Prospects for Sustainability*, Ashgate, UK: Aldershot.

McEldowney, J.F. and McEldowney, S. (1996) *Environment and the Law: An Introduction for Environmental Scientists and Lawyers*, Harlow, UK: Addison-Wesley Longman.

McGillivray, D. and Mansell, W. (1998) 'The Water of the Danube: the ICJ Bottles It', *Water Law*, vol. 9, no. 3, pp. 107–18.

M'Gonigle, R.M. and Zacher, M.W. (1979) *Pollution, Politics and International Law: Tankers at Sea*, Berkeley: University of California Press.

MacIntyre, A.C. (1985) *After Virtue: A Study in Moral Theory* (2nd edn), London: Duckworth.

McIntyre, O. and Mosedale, T. (1997) 'The Precautionary Principle as a Norm of Customary International Law', *Journal of Environmental Law*, vol. 9, no. 2, pp. 221–41.

McKay, S., Pearson, M. and Smith, S. (1990) 'Fiscal Instruments in Environmental Policy', *Fiscal Studies*, November, pp. 1–20.

McKibben, B. (1990) *The End of Nature*, New York: Doubleday.

Mackie, J.L. (1984) 'Can there be a Right-Based Moral Theory?' in J. Waldron (ed.) *Theories of Rights*, New York: Oxford University Press.

Mackintosh, J.P. (1982) *The Government and Politics of Britain*, London: Hutchinson.

McLaren, J.P.S. (1983) 'Nuisance Law and the Industrial Revolution: Some Lessons from Social History', *Journal of Legal Studies*, vol. 3, p. 155.

McLaughlin, A. (1993) *Regarding Nature: Industrialism and Deep Ecology*, New York: State University of New York Press.

McLeod, G. (1995) 'Do we need an Environmental Court in Britain?' in D. Robinson and J. Dunkley, *Public Interest Perspectives in Environmental Law*, London: Wiley Chancery.

McNair (1934) 'International Legislation', *Iowa Law Review*, vol. 19, p. 177.

Macrory, R. (1992) 'The Enforcement of Community Environmental Laws: Some Critical Issues', *Common Market Law Review*, vol. 29, pp. 347–69.

Maastricht University (http://www.unimaas.nl/~egmilieu/).

Majone, G. (1984), 'Science and Trans-science in Standard Setting', *Science, Technology and Human Values*, vol. 9, pp. 15–22.

Malthus, T. (1970, orig. 1798) *An Essay on the Principle of Population*, Harmondsworth, UK: Penguin.

Marcuse, H. (1964) *One-Dimensional Man*, London: Routledge.

Marietta, D.E. (1988) 'Environmental Holism and Individuals', *Environmental Ethics*, vol. 10, no. 3, pp. 251–8

Martínez Cobo, José R. (1987) *Study of the Problem of Discriminating Against Indigenous Populations*, vol. 5: *Conclusions: Proposals, Recommendations*, New York: United Nations.

Marx, K. (1875) 'Critique of the Gotha program' in R.C. Tucker (ed.) (1972) *The Marx-Engels Reader*, New York: Norton.

Matthews, P. (1996) 'Problems Related to the Convention on the International Trade in Endangered Species', *International and Comparative Law Quarterly*, vol. 45, April, pp. 421–31.

Mayr, E. (1969) 'The Biological Meaning of Species', *Biological Journal of the Linnean Society*, vol. 1, p. 311.

Mazza, M.A. (1996) 'The New Evidential Privilege for Environmental Audit Reports: Making the Worst of a Bad Situation', *Ecology Law Quarterly*, vol. 23, no. 1, p. 79.

Meadows, D.H. *et al.* (1974) *The Limits to Growth*, London: Pan Books.

Meadows, D.H., Meadows, D.L. and Randers, J. (1992) *Beyond the Limits: Global Collapse or a Sustainable Future*, London: Earthscan.

Meier, P.C. (1995) '*Stevens v. City of Cannon Beach*: Taking Takings into the Post-*Lucas* Era', *Ecology Law Review*, vol. 22, pp. 413–48.

Melke, J.E. (1990) *Oil in the Ocean: the Short and Long-Term Impacts of a Spill* (CRA Report for Congress), Washington, DC: Library of Congress.

Merchant, C. (1987) 'The Theoretical Structure of Ecological Revolutions', *Environmental Review*, vol. 11, no. 4, pp. 265–74.

Meyers, G.D. (1991) 'Old-Growth Forests, the Owl, and Yew: Environmental Ethics versus Traditional Dispute Resolution Under the Endangered Species Act and Other Public Lands and Resources Laws', *Environmental Affairs*, vol. 18, pp. 623–68.

Midgley, M. (1983) *Animals and Why they Matter*, Athens, Ga.: University of Georgia Press.

Miers, D.R. and Page, A.C. (1990) *Legislation*, (2nd edn), London: Sweet and Maxwell.

Mill, J.S. (1891) *Principles of Political Economy, with some of their Applications to Social Philosophy*, London: Routledge.

Mill, J.S. (1958) *Considerations on Representative Government*, Indianapolis: Bobbs-Merrill.

Mill, J.S. (1972, orig. 1863) *Utilitarianism, Liberty, Representative Government, with Selections from Auguste Comte and Positivism*, H.B. Acton (ed.), London: Dent.

Miller, J. (2000) 'The Standing of Citizens to Enforce against Violations of Environmental Statutes in the United States', *Journal of Environmental Law*, vol. 12, no. 3, pp. 370–81.

Milne, A. (1993) 'The Perils of Green Pessimism', *New Scientist*, vol. 138, no. 1877, 12 June, pp. 34–7.

Milton, K. (1991) 'Interpreting Environmental Policy: A Social Scientific Approach', *Journal of Law and Society*, vol. 18, no. 1, pp. 4–29.

Mishan, E.J. (1993) *The Costs of Economic Growth* (rev. edn), London: Weidenfeld and Nicolson.

Mitchell, R.B. (1996) 'Compliance Theory: An Overview' in J. Cameron and P. Roderick, *Improving Compliance with International Environmental Law*, London: Earthscan.

Moonshop (1999) 'Welcome to Moonshop' (http://www.moonshop.com/).

Moore, G.E. (1903) *Principia Ethica*, Cambridge: Cambridge University Press.

Moore, M.S. (1997) 'Legal Principles Revisited', *Iowa Law Review*, vol. 82, pp. 867–91.

Moore, V. (1987) *A Practical Approach to Planning Law*, London: Blackstone Press.

Morrow, K. and Turner, S. (1998) 'The More Things Change, The More they Stay the Same? Environmental Law, Policy and Funding in Northern Ireland', *Journal of Environmental Law*, vol. 10, no. 1, pp. 41–59.

Munn, K. (1999) 'Responsible Care and Related Voluntary Initiatives to Improve Enterprise Performance on Health, Safety and Environment in the Chemical Industry' (http://www.ilo.org/public/english/dialogue/sector/papers/respcare/rscare6.htm).

Munro, C. (1975) 'Laws and Conventions Distinguished', *Law Quarterly Review*, vol. 91, p. 218.

Myers, N. and Simon, J. (1994) *Scarcity or Abundance? A Debate on the Environment*, New York: W.W. Norton.

Naess, A. (1979) 'Self-realisation in Mixed Communities of Humans, Bears, Sheep and Wolves', *Inquiry*, vol. 22, pp. 231–42.

Naess, A. (1989) *Ecology, Community and Lifestyle: Outline of an Ecosophy*, Cambridge: Cambridge University Press.

Nanda, V.P. (1995a) *International Environmental Law and Policy*, New York: Transnational Publishers.

Nanda, V.P. (1995b) 'Environment' in O. Schacter and C.C. Joyner (eds) *United Nations Legal Order*, Cambridge: Grotius Publications.

NASA (1996) 'Meteorite Yields Evidence of Primitive Life on Early Mars' (http://www.jpl.nasa.gov/snc/nasa1.html).

Nash, R. (1967) *Wilderness and the American Mind*, New Haven, Conn.: Yale University Press.

Nash, R.F. (1989) *The Rights of Nature: A History of Environmental Ethics*, Madison: University of Wisconsin Press.

Nietzsche, F. ([1887] 1969) *On the Genealogy of Morals* (trans. W. Kaufman and R.J. Hollingdale), New York: Vintage Books.

Nollkaemper, A. (1991) 'The Precautionary Principle in International Environmental Law: What's New Under the Sun?', *Marine Pollution Bulletin*, vol. 22, no. 3, pp. 107–10.

North, R. (1995) *Life on a Modern Planet*, Manchester: Manchester University Press.

Northey, R. (1991) 'Federalism and Comprehensive Environmental Reform', *Osgoode Hall Law Journal*, vol. 29, p. 127.

Note (1979) 'Endangered Species Act Amendments of 1978: A Congressional Response to Tennessee Valley Authority v. Hill', *Columbia Journal of Environmental Law*, vol. 5, p. 283.

Note (1991) 'Lujan v. National Wildlife Federation: The Supreme Court Tightens the Reins on Standing for Environmental Groups', *Catholic University Law Review*, vol. 40, p. 443.

Note (1992) '"Ask a Silly Question . . .": Valuation of Natural Resource Damage', *Harvard Environmental Law Review*, vol. 105, p. 1981.

Nozick, R. (1974) *Anarchy, State and Utopia*, Oxford: Blackwell.

Odum, E., in collaboration with Odum, H. T. (1959) *Fundamentals of Ecology*, Philadelphia: Pa.: Saunders.

OECD (1975) *Council Recommendation on the Application of the Polluter Pays Principle*, reprinted in *International Legal Materials*, vol. 14, no. 138.

Ogus, A. (1994) 'Standard Setting for Environmental Protection: Principles and Processes' in M. Faure, J. Vervaele and A. Weale, *Environmental Standards in the European Union in an Interdisciplinary Framework*, Antwerp: Malku.

Ogus, A. and Richardson, G.M. (1977) 'Economics and the Environment: A Study of Private Nuisance', *Cambridge Law Journal*, vol. 36, no. 2, pp. 284–325.

Olowofoyeku, A. (1999) 'Decentralising the UK: the Federal Argument', *Edinburgh Law Review*, vol. 3, pp. 57–84.

Olson, D.M. (1980) *The Legislative Process: A Comparative Approach*, New York: Harper and Row.

O'Neill, O. (1991) 'Kantian Ethics' in P. Singer (ed.) *A Companion to Ethics*, Oxford: Blackwell.

O'Quinn, J.C. (2000) 'Not-So-Strict-Liability: A Foreseeability Test for *Rylands v. Fletcher* and other lessons from *Cambridge Water Co. v. Eastern Counties Leather plc', The Harvard Environmental Law Review*, vol. 24, no. 1, pp. 287–313.

O'Riordan, T. (1981) *Environmentalism* (2nd edn), London: Pion.

O'Riordan, T. and Jordan, A. (1995) 'The Precautionary Principle in Contemporary Environmental Politics', *Environmental Values*, vol. 4, p. 191.

Osborn, D. (1999) 'Setting Environmental Standards. Twenty-first Report by the Royal Commission on Environmental Pollution, 1998', *Journal of Environmental Law*, vol. 11, no. 1, pp. 218–20.

Paeffgen, H.U. (1991) 'Overlapping Tensions between Criminal and Administrative Law: The Experience of West German Environmental Law', *Journal of Environmental Law*, vol. 3, no. 2, pp. 247–64.

Pallaemerts, M. (1993) 'International Environmental Law from Stockholm to Rio: Back to the Future?', in P. Sands (ed.) *Greening International Law*, London: Earthscan.

Palmer, C. (1999) 'Education or Catastrophe? Environmental Values and Environmental Education' (http://www.farmington.ac.uk/library/papers/ei/ei2.html).

Palmer, G. (1992) 'New Ways to Make International Environmental Law', *American Journal of International Law*, vol. 86, pp. 259–83.

Palmer, G.W.R. (1989) 'Settlement of International Disputes: The "Rainbow Warrior" Affair', *Commonwealth Law Bulletin*, vol. 15, p. 585.

Palomares-Soler, M. and Thimme, P. (1996) 'Environmental Standards: EMAS and ISO 14001 Compared', *European Environmental Law Review*, vol. 5, nos. 8/9, pp. 247–51.

Pashukanis, E.B. (1978) *Law and Marxism: A General Theory*, trans. B. Einhorn, London: Ink Links.

Passmore, J. (1980) *Man's Responsibility for Nature: Ecological Problems and Western Tradition* (2nd edn), London: Duckworth.

Passmore, J. (1985) 'Attitudes to Nature' in R. Elliot (ed.) *Environmental Ethics*, Oxford: Oxford University Press.

Paton, H.J. (1948) *The Moral Law*, London: Hutchinson.

Pearce, D.W. and Turner, R.K. (1990) *Economics of Natural Resources and the Environment*, Hemel Hempstead, UK: Harvester Wheatsheaf.

Pelzer, N. (1988) 'Concepts of Nuclear Liability Revisited: A Post-Chernobyl Assessment of the Paris and the Vienna Conventions' in P. Cameron (ed.) *Nuclear Energy Law after Chernobyl*, London: Graham & Trotman.

Pence, G. (1991) 'Virtue Theory' in P. Singer (ed.) *A Companion to Ethics*, Blackwell: Oxford, and Cambridge, Mass.

Pepper, D. (1984) *The Roots of Modern Environmentalism*, London: Routledge.
Pepper, D. (1993) *Eco-socialism*, London: Routledge.
Permanent.com (1999) 'Legal Histories and Issues' (http://permanent.com/ep_legal.htm).
Perrins, C. (1994) 'Sustainable Livelihoods and Environmentally Sound Technologies', *International Labour Review*, vol. 133, no. 3, pp. 305–26.
Perry, S.R. (1997) 'Two Models of Legal Principles', *Iowa Law Review*, vol. 82, pp. 787–819.
Pezzey, J. (1988) 'Market Mechanisms of Pollution Control' in K. Turner (ed.) *Sustainable Environmental Management: Principles and Practice*, London: Belhaven Press.
Phipps, E.W. and Beck, H.U. (1997) 'United Nations Institute for Training and Research', *Yearbook of International Environmental Law*, vol. 8, pp. 563–70.
Pickett, S.T.A and White, P.S. (1985) *The Ecology of Natural Disturbances and Patch Dynamics*, Orlando, Fl.: Academic Press.
Pinchot, G. (1914) *The Training of a Forester*, Philadelphia, Pa.: Lippincott.
Piradov, A.S. (1976) *International Space Law*, Moscow.
Pluhar, E.B. (1995) *Beyond Prejudice: The Moral Significance of Human and Nonhuman Animals*, Durham, N.C.: Duke University Press.
Plumwood, V. (1995) 'Has Democracy Failed Ecology; An Ecofeminist Perspective', *Environmental Politics*, vol. 4, no. 4, pp. 134–68.
Pocklington, D. (1997) *The Law of Waste Management*, Crayford, UK: Shaw and Sons.
Pontin, B. (1999) 'Modernising Local Government: Local Democracy and Community Leadership', *Environmental Law and Management*, vol. 10, no. 4, pp. 177–9.
Popper, F.R. (1934) *The Logic of Scientific Discovery*, London: Hutchinson.
Popper, F.R. (1966) *The Open Society and its Enemies*, London: Routledge.
Porras, I.M. (1993) 'The Rio Declaration: A New Basis for International Cooperation', in P. Sands (ed.) *Greening International Law*, London: Earthscan.
Posner, R.A. (1972) 'A Theory of Negligence', *Journal of Legal Studies*, vol. 1, p. 29.
Poustie, M. (1999) 'The Scotland Act 1998', *Environmental Law and Management*, vol. 11, pts 1/2, pp. 8–10.
Rabie, M.A. and Loubser, M.M. (1990) 'Legal Aspects of Weather Modification', *The Journal of Comparative and International Law*, vol. 23, no. 2, p. 177.
Rabin, R.L. (1987) 'Environmental Liability and the Tort System', *Houston Law Review,* vol. 24, pp. 27–52.
Rackham, O. (1976) *Trees and Woodland in the British Landscape*, London: Dent.
Radbruch, G. ([1946] 1990) 'Gesetzliches Unrecht und übergesetzliches Recht', in W. Hassemer (ed.) *Gustav Radbruch Gesamtausgabe*, vol. 3, Heidelberg: C.F. Müller.

Ramaekers-Jorgensen, D.A. (1999) 'Coping with Scientific Uncertainty in EU Environmental Legislation', *Environmental Liability*, vol. 7, no. 2, pp. 44–59.

Rawls, J. (1972) *A Theory of Justice*, Oxford: Clarendon Press.

Raz, J. (1990) *Practical Reason and Norms*, Princeton, N.J.: Princeton University Press.

Reeds, J. (1994) 'The Hazards of Self-Control', *The Surveyor*, 5 May, pp. 12–15.

Regan, T. (1983) *The Case for Animal Rights*, London: Routledge and Kegan Paul.

Rehbinder, E. (1996) 'Self-regulation by Industry' in G. Winter, *European Environmental Law: A Comparative Perspective*, Aldershot: Dartmouth.

Reichard, D. and McGarrity, J. (1994) 'Early Adolescents' Perceptions of Relative Risk from 10 Societal and Environmental Hazards', *Journal of Environmental Education*, vol. 26, no. 1, p. 16.

Reid, C.T. (ed.) ([1992] 1997), *Green's Guide to Environmental Law in Scotland*, Edinburgh: W. Green.

Rescher, N. (1980) 'Why Save Endangered Species?' in *Unpopular Essays on Technological Progress*, Pittsburgh, Pa.: University of Pittsburgh Press.

Rest, A. (1998) 'The Indispensability of an International Environmental Court', *Review of European Community and International Environmental Law*, vol. 7, no. 1, pp. 63–7.

Richardson, B. (1998) 'Economic Instruments and Sustainable Management in New Zealand', *Journal of Environmental Law*, vol. 10, pp. 29–39.

Richardson, G., Ogus, A. and Burrows, P. (1982) *Policing Pollution: A Study of Regulation and Enforcement*, Oxford: Clarendon Press.

Richardson, J. (ed.) (1982) *Policy Styles in Western Europe*, London: Allen and Unwin.

Ricketts, M.J. and Peacock, A.T. (1984) *The Regulation Game: How British and West German Companies Bargain with Government*, Oxford: Blackwell.

Rielly, B. (1996) 'Clear and Present Danger: A Role for the United Nations Security Council in Protecting the Global Environment', *Melbourne Law Review*, vol. 20, pp. 763–804.

Roberts, L.D. (1997) 'Environmental Standards for Prospective Non-Terrestrial Development' (http://www.permanent.com/archimedes/EnvironmentArticle.html).

Roberts, T.C.L. (1987) 'Allocation of Liability under CERCLA: A Carrot and Stick Formula', *Ecology Law Quarterly*, vol. 14, pt 4, pp. 601–38.

Robinson, D. (1995) 'Public Interest Environmental Law – commentary and analysis' in D. Robinson and J. Dunkley, *Public Interest Perspectives in Environmental Law*, London: Wiley Chancery.

Rogers, C.R. (1961) *On Becoming a Person: A Therapist's View of Psychotherapy*, Boston, Mass.: Houghton Mifflin.

Roht-Arriaza, N. (1995) 'Shifting the Point of Regulation: The International Organization for Standardization and Global Lawmaking on Trade and the Environment', *Ecology Law Quarterly*, vol. 22, no. 3, p. 479.

Rolston, H. (1988) *Environmental Ethics: Duties to and Values in the Natural World*, Philadelphia, Pa.: Temple University Press.

Rolston, H. (1997) 'Nature for Real: Is Nature a Social Construct' in T.D.J. Chappell, *The Philosophy of the Environment*, Edinburgh: Edinburgh University Press.

Rose, C.M. (1985) 'Possession as the Origin of Property', *The University of Chicago Law Review*, vol. 52, pp. 73–88.

Rose, G. (1995) 'Regional Environmental Law in South East Asia', *Review of European Community and International Environmental Law*, vol. 4, no. 1, pp. 40–8.

Rose-Ackerman, S. (1977) 'Market Models for Water Pollution Control: Their Strengths and Weaknesses', *Public Policy*, vol. 25, no. 3, pp. 383–406.

Rothwell, D.R. and Boer, B. (1998) 'The Influence of International Environmental Law on Australian Courts', *Review of European Community and International Environmental Law*, vol. 7, no. 1, pp. 31–9.

Royal Commission on Environmental Pollution (1994) *Freshwater Quality: Sixteenth Report*, London: HMSO.

Royal Commission on Environmental Pollution (1998) *Setting Standards: Twenty-First Report by the Royal Commission on Environmental Pollution*, Cm 4053, London: HMSO.

Sagoff, M. (1981) 'Do we Need a Land Use Ethic?', *Environmental Ethics*, vol. 3, pp. 293–308.

Sagoff, M. (1984) 'Animal Liberation and Environmental Ethics: Bad Marriage, Quick Divorce', *Osgoode Hall Law Journal*, vol. 22, no. 2, pp. 297–307.

Sagoff, M. (1988) 'Some Problems with Environmental Economics', *Environmental Ethics*, vol. 10, no. 1, pp. 55–74.

Sagoff, M. (1995) 'Can Environmentalists be Liberals?' in R. Elliot (ed.) *Environmental Ethics*, Oxford: Oxford University Press.

Salter, J. (1993) 'Environmental Standards and Testing', *European Environmental Law Review*, vol. 2, no. 10, pp. 276–83.

Sampford, C.J.G. (1990) 'Responsible Government and the Logic of Federalism: An Australian Paradox?', *Public Law*, pp. 90–115.

Sandhovel, A. (1998) 'What can be Achieved Using Instruments of Self-regulation in Environmental Policy Making?', *European Environmental Law Review*, vol. 7, pt 3, pp. 83–4.

Sands, P. (1995) *Principles of International Environmental Law*: vol. I, *Frameworks, Standards and Implementation*, Manchester: Manchester University Press.

Sands, P. (1996) in H. Somsen (ed.) *Protecting the European Environment, Enforcing E.C. Environmental Law*, London: Blackstone Press.

Sands, P. (1998) 'International Environmental Litigation: What Future', *Review of European Community and International Environmental Law*, vol. 7, pp. 1–3.

Sarkar, S. (1999) *Eco-socialism or Eco-Capitalism: A Critical Analysis of Humanity's Fundamental Choices*, London: Zed Books.

Sax, J. (1970) 'The Public Trust Doctrine in Natural Resources Law: Effective Judicial Implementation', *Michigan Law Review*, vol. 68, p. 471.

Sax, J. and Conner, R.L. (1972) 'Michigan's Environmental Protection Act of 1970: A Progress Report', *Michigan Law Review*, vol. 70, p. 1004.

Scheberle, D. (1994) 'Radon and Asbestos: A Study of Agenda Setting and Causal Stories', *Policy Studies Journal*, vol. 22, pp. 74–86.

Schikhof, S. (1998) 'Direct and Individual Concern in Environmental Cases: The Barriers to Prospective Litigants', *European Environmental Law Review*, October, pp. 276–81.

Schindler, D.W. (1987) 'Is the Whole Really More Than the Sum of the Parts?' in G.E. Likens *et al.* (eds) *Status and Future of Ecosystem Science*, New York: Institute of Ecosystem Studies.

Schneiderman, D. (1996) 'NAFTA's Takings Rule: American Constitutionalism Comes to Canada', *The University of Toronto Law Journal*, vol. 46, no. 4, p. 499.

Schrader-Frechette, K. and McCoy, E.D. (1994) 'How the Tail Wags the Dog: How Value Judgments Determine Ecological Science', *Environmental Values*, vol. 3, no. 2, pp. 107–20.

Schumacher, E. (1973) *Small is Beautiful*, London: Sphere.

Schwabach, A. (1989) 'The Sandoz Oil Spill: the Failure of International Law to Protect the Rhine from Pollution', *Ecology Law Quarterly*, vol. 16, pp. 443–80.

Schweitzer, A. (1923) *Civilization and Ethics*, trans. J.P. Naish, London: A&C Black Ltd.

Select Committee on Environment, Transport and Regional Affairs (1999) 'Thirteenth Report: The Planning Inspectorate and Public Inquiries – Future of the Planning Inspectorate' (http://www.publications.parliament.uk/pa/cm199900/cmselect/cmenvtra/364/36411.htm#a21).

Shelbourne, C. (2000) 'Case Law Analysis. Can the Insolvent Polluter Pay? Environmental Licences and the Insolvent Company. Official Receiver as Liquidator of Celtic Extraction Ltd and Bluestone Chemicals Ltd v Environment Agency', *Journal of Environmental Law*, vol. 12, no. 2, pp. 207–29.

Shere, M.E. (1995) 'The Myth of Meaningful Environmental Risk Assessment', *Harvard Environmental Law Review*, vol. 19, no. 2, pp. 409–92.

Shover, N., Lynxwiler, J., Groce, S. and Clelland, D. (1984) 'Regional Variation in Regulatory Law Enforcement: the Surface Mining Control and Reclamation Act of 1977' in K. Hawkins and J.M. Thomas (eds) *Enforcing Regulation*, The Hague: Kluwer-Nijhoff.

Silver, L. (1986) 'The Common Law of Environmental Risk and Some Recent Applications', *Harvard Environmental Law Review*, vol. 10, p. 61.

Silver, L. (1999) *Remaking Eden: Cloning and Beyond in a Brave New World*, London: Weidenfeld and Nicolson.

Sinclair, D. (1997) 'Self-regulation versus Command and Control? Beyond False Dichotomies', *Law and Policy*, vol. 19, no. 4, pp. 529–59.

Singer, P. (1976) *Animal Liberation: A New Ethics for Our Treatment of Animals*, London: Cape.

Singh, N. (1987) 'Right to Environment and Sustainable Development as a Principle of International Law', *Journal of the Indian Law Institute*, vol. 29, no. 3, pp. 289–320.

Slaughter, A.M. (1995) 'International Law in a World of Liberal States', *European Journal of International Law*, vol. 6, no. 4, pp. 503–8.

Smart, J.J.C. and Williams, B. (1973) *Utilitarianism: For and Against*, Cambridge: Cambridge University Press.

Smets, H. (1994) 'The Polluter Pays Principle in the Early 1990s', in L. Campiglio, L. Pineschi, D. Siniscalco and T. Treves (eds) *The Environment After Rio: International Law and Economics*, London: Graham and Trotman.

Smith, C., Collar, N. and Poustie, M. (1997) *Pollution Control: The Law in Scotland* (2nd edn), Edinburgh: T & T Clark.

Sober, E. (1995) 'Philosophical Problems for Environmentalism' in R. Elliot, *Environmental Ethics*, Oxford: Oxford University Press.

Sohn, L.B. (1973) 'The Stockholm Declaration on the Human Environment', *Harvard International Law Journal*, vol. 14, p. 423.

Somsen, H. and Sprokkereef, A. (1996) 'Making Subsidiarity Work for the Environmental Policy of the European Community: The Role of Science', *International Journal of Bioscience and the Law*, vol. 1, pp. 37–67.

Sperling, K. (1997) 'Going Down the Takings Path: Private Property Rights and Public Interest in Land Use Decision-Making', *Environmental and Planning Law Journal*, vol. 14, no. 6, p. 427.

Stavenhagen, R. (1990) *The Ethnic Question: Conflicts, Development, and Human Rights*, Hong Kong: United Nations University Press.

Steiner, P. (1995) 'A Specialist Environmental Court: An Australian Perspective' in D. Robinson and J. Dunkley, *Public Interest Perspectives in Environmental Law*, London: Wiley Chancery.

Steinzor, R.I. (1996) 'Regulatory Reinvention and Project XL: Does the Emperor Have Any Clothes?', *Environmental Law Reporter*, vol. 26, p. 10,527.

Steinzor, R.I. (1998) 'Reinventing Environmental Regulation: The Dangerous Journey from Command to Self Control', *Harvard Environmental Law Review*, vol. 22, pp. 103–202.

Stewart, R.B. (1975) 'The Reformation of American Administrative Law', *Harvard Environmental Law Review*, vol. 88, pp. 1667–1813.

Stewart, R. (1977) 'Pyramids of Sacrifice? Problems of Federalism in Mandating State Implementation of National Environmental Policy', *Yale Law Journal*, vol. 86, pp. 1196–1212.

Stone, C. (1972) 'Should Trees Have Standing? – Toward Legal Rights for Natural Objects', *Southern California Law Review*, vol. 45, pp. 450–501.

Stone, C. (1985) 'Should Trees Have Standing? Revisited. How Far Will Law and Morals Reach – A Pluralist Perspective', *Southern California Law Review*, vol. 59, no. 1, pp. 1–154.

Stradling, D. (1999) 'The Smoke of Great Cities: British and American Efforts to Control Pollution, 1860–1914', *Environmental History*, vol. 4, no. 1, pp. 6–31.

Sugameli, G. P. (1999), 'Lucas v. South Carolina Coastal Council: The Categorical and Other "Exceptions" to Liability for Fifth Amendment Takings of Private Property Far Outweigh the "Rule"', *Environmental Law*, vol. 29, no. 4, p. 939.

Sumikura, I. (1998) 'A Brief History of Japanese Environmental Administration: a qualified success story?', *Journal of Environmental Law*, vol. 10, no. 2, pp. 241–56.

Sumner, L.W. (1981) *Abortion and Moral Theory*, Princeton: Princeton University Press.

Swedish Ministry of the Environment (1996) *Swedish Environmental Legislation, Booklets 1–4*, Stockholm: Cabinet Office.

Tansley, A. (1935) 'The Uses and Abuses of Vegetational Concepts', *Ecology*, vol. 16, pp. 284–307.

Tarlock, D.A. (1996) 'Environmental Law: Ethics or Science?', *Duke Environmental Law and Policy Forum*, vol. 7, no. 1, pp. 193–223.

Taubenfield, H. (ed.) (1968) *Weather Modification and the Law*, Dobbs Ferry, N.Y.: Oceana Publications.

Taylor, A. (1998) 'Wasty Ways: Stories of American Settlement', *Environmental History*, vol. 3, pt 3, pp. 291–310.

Taylor, P.W. (1986) *Respect for Nature: A Theory of Environmental Ethics*, Princeton, N.J.: Princeton University Press.

Thomas, K. (1984), *Man and the Natural World*, London: Penguin Books.

Thompson, B. (1993) 'Economic Integration Efforts in Africa – the Abuja Treaty', *African Journal of International and Comparative Law*, vol. 5, pt 4, pp. 743–67.

Thornton, J. and Beckwith, S. (1997) *Environmental Law*, London: Sweet and Maxwell.

Thornton, J. and Tromans, S. (1999) 'Human Rights and Environmental Wrongs: Incorporating the European Convention on Human Rights: Some Thoughts on the Consequences for UK Environmental Law', *Journal of Environmental Law*, vol. 11, no. 1, pp. 35–57.

Tietenberg, T.H. (1990) 'Economic Instruments for Environmental Regulation', *Oxford Review of Economic Policy*, vol. 6, no. 1, pp. 17–31.

von Tigerstrom, B. (1998) 'The Public Trust Doctrine in Canada', *Journal of Environmental Law and Practice*, vol. 7, no. 3, p. 379.

Toubes Muñiz, J.R. (1997) 'Legal Principles and Legal Theory', *Ratio Juris*, vol. 10, no. 3, pp. 267–87.

Trainer, F.E. (1985) *Abandon Affluence!*, London: Zed Books.

Tribe, L. (1974) 'Ways Not to Think about Plastic Trees', *The Yale Law Journal*, vol. 83, pp. 1315–48.

Tromans, S. (1995) 'High Talk and Low Cunning: Putting Environmental Principles into Legal Practice', *Journal of Planning and Environment Law*, September, pp. 779–96.

Turpin, C. (1985) *British Government and the Constitution: Text, Cases and Materials*, London: Weidenfeld and Nicolson.

Unabomber (1997) 'The Unabomber's Manifesto' (http://www.soci.niu.edu/~critcrim/uni/uni.html).

Underwood, E.A. (1948) 'History of Cholera in Great Britain', *Proceedings of the Royal Society of Medicine*, vol. 41, pp. 165–73.

Van der Wal, K. (1994) 'Technology and the Ecological Crisis' in W. Zweers (ed.) *Ecology, Technology and Culture*, Cambridge: White Horse Press.

vanDeVeer, D. (1979) 'Interspecific Justice', *Inquiry*, vol. 22, pp. 55–79.

Van Hoecke, M. (1995) 'The Use of Unwritten Legal Principles by Courts', *Ratio Juris*, vol. 8, no. 3, pp. 248–60.

Victor, D.G., Raustiala, K. and Skolnikoff, E.B. (eds) (1998) *The Implementation and Effectiveness of International Environmental Commitments*, Cambridge, Mass.: MIT Press.

Vogel, D. (1986) *National Styles of Regulation*, Ithaca, N.Y.: Cornell University Press.

Vogel, S. (1996) *The Concept of Nature in Critical Theory*, New York: State University of New York Press.

von Molkte, K. (1988) 'The *Vorsorgeprinzip* in West German Policy', in Royal Commission on Environmental Pollution, 12th Report, *Best Practicable Environmental Option*.

Wagner, W.E. (1999) 'Congress, Science and Environmental Policy', *University of Illinois Law Review*, pp. 181–286.

Walker, V.R. (1995) 'Direct Inference, Probability, and a Conceptual Gulf in Risk Communication', *Risk Analysis*, vol. 15, no. 5, p. 603.

Ware, E.F. (1905) *Roman Water Law*, St Paul, Minn.: West Publishing.

Warnock, G.J. (1971) *The Object of Morality*, London: Methuen.

Weiner, J.B. (1995) 'Review Essay, Law and the New Ecology: Evolution, Categories, and Consequences', *Ecology Law Quarterly*, vol. 22, pp. 325–57.

Weir, J. (1989) 'Species Extinction and the Concho Water Snake: a Case Study in Environmental Ethics', *Contemporary Philosophy*, vol. 12, no. 7, pp. 1–9.

Wenz, P.S. (1988) *Environmental Justice*, New York: State University of New York Press.

Werksman, J.C.J. and Roderick, P. (1996) *Improving Compliance with International Environmental Law*, London: Earthscan.

Wexler, J. (1987) 'The Rainbow Warrior Affair: State and Agent Responsibility for Authorized Violations of International Law', *Boston University International Law Journal*, vol. 5, p. 389.

Wildavsky, A. (1995) *But is it True? A Citizen's Guide to Environmental Health and Safety Issues*, Cambridge, Mass.: Harvard University Press.

Wilkinson, D. (1992) 'Maastricht and the Environment: the Implications for the EC's Environmental Policy of the Treaty on European Union', *Journal of Environmental Law*, vol. 4, pp. 221–31.

Wilkinson, D. (1993) 'Moving the Boundaries of Compensable Environmental Damage: the Effect of Two New International Protocols', *Journal of Environmental Law*, vol. 5, no. 1, pp. 71–90.

Wilkinson, D. (1994) 'Cambridge Water Company v Eastern Counties Leather plc: Diluting Liability for Continuing Escapes', *Modern Law Review*, vol. 57, pp. 799–811.

Wilkinson, D. (1997) 'An Idiomatic Discussion of Environmental Legislation' in P. Ireland and P. Laleng, *The Critical Lawyers' Handbook 2*, London: Pluto Press.

Wilkinson, D. (1999) 'Using Environmental Ethics to Create Ecological Law' in J. Holder and D. McGillivray (eds) *Locality and Identity: Environmental Issues in Law and Society*, Dartmouth: Aldershot.

Williams, D.A.R. and Grinlinton, D. (1997) 'Introduction to Environmental Law' in D.A.R. Williams, *Environmental and Resource Management Law in New Zealand*, Wellington, NZ: Butterworths.

Williams, P.R. (1995) 'Can International Legal Principles play a Positive Role in Resolving Central and Eastern European Transboundary Environmental Disputes?', *Georgetown International Environmental Law Review*, vol. 7, pt 2, pp. 421–62.

Wils, W.P.J. (1994) 'Subsidiarity and EC Environmental Policies', *Journal of Environmental Law*, vol. 6, pt 1, pp. 85–91.

Wilson, E.O. (1992) *The Diversity of Life*, London/New York: Penguin.

Wilson, W. (1999) *Making Environmental Laws Work: Law and Policy in the UK and USA*, Oxford: Hart Publishing.

Winpenny, J.T. (1991) *Values for the Environment: A Guide to Economic Assessment*, London: HMSO.

Winter, G. (1985) 'Bartering Rationality in Regulation', *Law and Society Review*, vol. 19, pp. 219–50.

Winter, G. (1994) *German Environmental Law: Basic Texts and Introduction*, Dordrecht/London: Martinus Nijhoff/Graham and Trotman.

Winter, G. (1996a) 'Standard-setting in Environmental Law' in G. Winter (ed.) *European Environmental Law: A Comparative Perspective*, Aldershot: Dartmouth.

Winter, G. (1996b) 'Freedom of Environmental Information' in G. Winter, *European Environmental Law: a Comparative Perspective*, Aldershot: Dartmouth.

Woodbury, S.E. (1987) 'Aesthetic Nuisance: The Time has Come to Recognise It', *Natural Resources Journal*, vol. 27, pt 4, pp. 877–86.

Woolf, H. (1992) 'Are the Judiciary Environmentally Myopic?', *Journal of Environmental Law*, vol. 4, no. 1, p. 1.

World Affairs Council of Philadelphia (1975) 'A Declaration of Interdependence' (http://www.wfa.org/readings/wrldaff.htm).

World Commission on Environment and Development (WCED) (1987) *Our Common Future*, New York: Oxford University Press.

The World Socialist Party (2000) 'Imagine a World without Laws', *Socialist Standard*, November (http://www.worldsocialism.org/spgb/nov00/law.html).

WWF (1994) 'Introduction to the Law of the Sea' (http://www.clark.net/pub/diplonet/los_guide.html).

Yandle, B. (1997) *Common Sense and Common Law for the Environment: Creating Wealth in Hummingbird Economies*, Lanham, Md.: Rowman and Littlefield.

Yeager, P. (1999) 'Structural Bias in Regulatory Law Enforcement: The Case of the U.S. Environmental Protection Agency' in B. Hutter (ed.) *A Reader in Environmental Law*, Oxford: Oxford University Press.

Yearly, S. (1991) *The Green Case: A Sociology of Environmental Issues, Arguments and Politics*, London: HarperCollins.

Yearly, S. (1992) ' Green Ambivalence about Science: Legal-rational Authority and the Scientific Legitimation of a Social Movement', *British Journal of Sociology*, vol. 43, p. 511.

Young, S. (1991) 'Regulatory and Judicial Responses to the Possibility of Biological Hazards from Electromagnetic Fields Generated by Power Lines', *Villanova Law Review*, vol. 36, pp. 129–90.

Yuhuke, R. (1994) 'Launch of a Movement', *The Environmental Forum*, vol. 11, pt 6, pp. 32–40.

Zander, M. (1980) *The Law-making Process*, London: Weidenfeld and Nicolson.

Index